Aquatic organisms swim in a variety of ways,   ction; they swim at a wide range of speeds and span a vast size range, from bacteria to protists, to the largest whales. One of the most fascinating aspects of aquatic locomotion is the remarkable sets of adaptions that have been evolved for different purposes.

This volume brings together current research on a wide range of swimming organisms, with an emphasis on the biomechanics, physiology and hydrodynamics of swimming in or on water. Several chapters deal with different aspects of fish swimming, from the use of different 'gaits' to the operation of the locomotor muscles. All chapters are by recognised authorities in their different fields, and all are accessible to biologists interested in aquatic locomotion.

MECHANICS AND PHYSIOLOGY OF ANIMAL SWIMMING

MECHANICS AND PHYSIOLOGY
OF ANIMAL SWIMMING

Edited by

LINDA MADDOCK

Marine Biological Association

QUENTIN BONE

Marine Biological Association

JEREMY M.V. RAYNER

School of Biological Sciences, University of Bristol

A volume arising from the Symposium on **Mechanics and Physiology
of Animal Swimming**, organized by the Marine Biological Association of the
United Kingdom and the Society for Experimental Biology and held at the
Polytechnic South West, Plymouth, 15-18 April 1991

CAMBRIDGE
UNIVERSITY PRESS

CAMBRIDGE UNIVERSITY PRESS
Cambridge, New York, Melbourne, Madrid, Cape Town, Singapore, São Paulo

Cambridge University Press
The Edinburgh Building, Cambridge CB2 8RU, UK

Published in the United States of America by Cambridge University Press, New York

www.cambridge.org
Information on this title: www.cambridge.org/9780521460781

First published 1994
This digitally printed version 2008

A catalogue record for this publication is available from the British Library

ISBN 978-0-521-46078-1 hardback
ISBN 978-0-521-06495-8 paperback

Contents

Contributors

J.D. Altringham, Department of Pure and Applied Biology, University of Leeds, Leeds, West Yorkshire, LS2 9JT, UK

R. Bannasch, TU-Berlin, Bionik und Evolutionstechnik, Ackerstraße 71-76, D-1000 Berlin 65, Germany

Q. Bone, Marine Biological Association, The Laboratory, Citadel Hill, Plymouth, PL1 2PB, UK

G. Bowtell, Department of Mathematics, City University, Northampton Square, London, EC1V 0HB, UK

R.W. Brill, National Marine Fisheries Service, Honolulu, Hawaii 96822, USA

J.C. Carling, Department of Physiology, St George's Hospital Medical School, University of London, Cranmer Terrace, Tooting, London, SW17 0RE, UK

H. Dewar, Center for Marine Biotechnology and Biomedicine, Scripps Institution of Oceanography, University of California at San Diego, La Jolla, California 92093-0204, USA

J. Febvre, Université de Nice-Sophia-Antipolis, 06108 Nice Cedex, France

C. Febvre-Chevalier, URA 671 CNRS, Observatoire Océanologique, 06230 Villefranche-sur-Mer, France

P.A. Fields, Center for Marine Biotechnology and Biomedicine, Scripps Institution of Oceanography, University of California at San Diego, La Jolla, California 92093-0204, USA

F.E. Fish, Department of Biology, West Chester University, West Chester, Pennsylvania 19383, USA

J.B. Graham, Center for Marine Biotechnology and Biomedicine, Scripps Institution of Oceanography, University of California at San Diego, La Jolla, California 92093-0204, USA

N.A. Hill, Department of Applied Mathematical Studies, The University, Leeds, LS2 9JT, UK

J.A. Hoar, Biology Department and Aquatron Laboratory, Dalhousie University, Halifax, Nova Scotia, B3H 4J1, Canada

M.A. Hoelzer, Department of Biology, The University of Alaska, Anchorage, Alaska 99508, USA

J.O. Kessler, Department of Physics, The University of Arizona, 1118 E 4th Street, Tucson, Arizona 85721, USA

T. Knower, Center for Marine Biotechnology and Biomedicine, Scripps Institution of Oceanography, University of California at San Diego, La Jolla, California 92093-0204, USA

K.E. Korsmeyer, Center for Marine Biotechnology and Biomedicine, Scripps Institution of Oceanography, University of California at San Diego, La Jolla, California 92093-0204, USA

N.C. Lai, Center for Marine Biotechnology and Biomedicine, Scripps Institution of Oceanography, University of California at San Diego, La Jolla, California 92093-0204, USA

L. Maddock, Marine Biological Association, The Laboratory, Citadel Hill, Plymouth, PL1 2PB, UK

J.A. Massare, Department of Earth Sciences, State University of New York College at Brockport, Brockport, New York 14420, USA

R.K. O'Dor, Biology Department and Aquatron Laboratory, Dalhousie University, Halifax, Nova Scotia, B3H 4J1, Canada

T.J. Pedley, Department of Applied Mathematical Studies, The University, Leeds, LS2 9JT, UK

J.M.V. Rayner, School of Biological Sciences, University of Bristol, Woodland Road, Bristol, BS8 1UG, UK

L.C. Rome, Department of Biology, Leidy Laboratory, University of Pennsylvania, Philadelphia, Pennsylvania 19104-6018, USA

R. Shabetai, Department of Medicine, UCSD and Veterans Administration Medical Center, San Diego, California 92161, USA

R.E. Shadwick, Center for Marine Biotechnology and Biomedicine, Scripps Institution of Oceanography, University of California at San Diego, La Jolla, California 92093-0204, USA

E. Sim, Biology Department and Aquatron Laboratory, Dalhousie University, Halifax, Nova Scotia, B3H 4J1, Canada

M.A. Taylor, Department of Geology, National Museums of Scotland, Chambers Street, Edinburgh, EH1 1JF, UK

J.J. Videler, Department of Marine Biology, University of Groningen, PO Box 14, 9750 AA, Haren, The Netherlands

C.S. Wardle, SOAFD Marine Laboratory, PO Box 101, Victoria Road, Aberdeen, AB9 8DB, UK

P.W. Webb, School of Natural Resources & Environment, University of Michigan, Ann Arbor, Michigan 48109-1115, USA

D.M. Webber, Biology Department and Aquatron Laboratory, Dalhousie University, Halifax, Nova Scotia, B3H 4J1, Canada

T.L. Williams, Department of Physiology, St George's Hospital Medical School, University of London, Cranmer Terrace, Tooting, London, SW17 0RE, UK

Introduction

Quentin Bone and Linda Maddock

Marine Biological Association, The Laboratory, Citadel Hill,
Plymouth, PL1 2PB

Aquatic organisms swim in a variety of ways, from jet propulsion to ciliary action; they swim at a wide range of speeds and span a vast size range, from bacteria and protists, to the largest whales. In consequence of the enormous size and speed range of swimming organisms, they operate under notably different Reynolds number regimes. This has led to very different selection pressures in different forms, and one of the fascinating aspects of aquatic locomotion is the remarkable sets of adaptations that have been evolved for different purposes. These are seen not only in external form, as in the body shapes of fish, penguins, and fossil marine reptiles, but also in the way the locomotor muscles are designed and controlled, and in the structure of the skeleton.

The different chapters consider some of the problems faced by swimmers from several points in the vast array of aquatic organisms, from the biological and physical effects determining the patterns of bacterial populations, to the remarkable special adaptations of penguins for underwater 'flight'. All organisms are constructed from materials that are denser than the water in which they swim. To avoid sinking either they must use buoyancy strategies, such as changing in shape, as do some marine protists, or storing light materials (like fat or gas) to provide static lift, or they must generate dynamic lift, as ichthyosaurs apparently did. Most fishes and all aquatic tetrapods have gas-filled swim bladders or lungs and hence are close to neutral buoyancy (at least at a particular depth). Some tetrapods, however, like penguins and plesiosaurs, can apparently achieve neutral buoyancy only by swallowing stones as ballast.

Although much can be inferred about the modes of life of fossil aquatic reptiles by consideration of living aquatic tetrapods, they are regrettably unavailable for observation and experiment, and it is with fishes (and their competitors, the squids) that most is known of the operation and control of the locomotor systems.

Squid are renowned for their jet propulsion, but their fins play an important and much less well known role in swimming. Like terrestrial animals, fish can move at different

Maddock, L., Bone, Q. & Rayner, J.M.V. (ed.). *Mechanics and Physiology of Animal Swimming.*
© 1994. Cambridge University Press.

speeds, and the introduction of the concept of gaits to fish swimming (from terrestrial animal locomotion studies), provides a new approach to fish swimming. The great majority of fish swim by oscillating the body with the myotomal muscles, and there has been considerable interest recently in the way that the control and timing of contraction of these segmental muscles is brought about. This has been examined in detail theoretically and experimentally, as have the concomitant physiological adaptations of the circulatory and respiratory systems in fast-swimming fish. Earlier experimental work on the physiology of swimming in fish involved small fish in relatively small flumes or tunnel respirometers. The development of a large portable high-speed water tunnel respirometer has enabled this approach to be extended to the most active fish, such as tunas and sharks.

Not all aspects of the biology of aquatic animal locomotion are considered in this book. The chapters range from bacteria and protist swimming and buoyancy, to underwater 'flight' by penguins, and formation swimming on the surface by ducklings, via squid, fish and fossil reptiles, and the reader will find a wide variety of approaches to the problems, common to all aquatic organisms, of moving through water and avoiding sinking.

This volume has chapters from many of the contributors to a symposium meeting with the same title, held at Plymouth in 1991, under the joint auspices of the Marine Biological Association of the United Kingdom, and the Society for Experimental Biology. Quentin Bone is grateful to the Leverhulme Trust for the award of an Emeritus Fellowship, during the tenure of which he edited this book. We thank all those who contributed to the success of the original meeting, and to the production of this volume. Tracie Endicott and Adrian Bonsey provided invaluable assistance with the editing and preparation of the camera-ready copy.

Chapter 1

Functional patterns of swimming bacteria

J.O. KESSLER*†, M.A. HOELZER‡, T.J. PEDLEY* AND N.A. HILL*

*Department of Applied Mathematical Studies, The University, Leeds, LS2 9JT, UK
†Department of Physics, University of Arizona, Tucson, AZ 85721, USA
‡Department of Biology, The University of Alaska, Anchorage, AK 99508, USA

Concentrated populations of the aerobic swimming bacteria *Bacillus subtilis* rapidly use up the oxygen dissolved in their culture medium. As a result, oxygen diffuses in from the air interface, creating an upward concentration gradient. The organisms then swim towards the surface where they accumulate. Because this arrangement of mass density is unstable, the entire fluid culture convects. Biological and physical factors thus jointly serve to organize the population, yielding dynamics which greatly improve the transport and mixing of oxygen and the viability of the cells.

INTRODUCTION

Concentrated fluid cultures of swimming bacterial cells, such as motile strains of *Bacillus subtilis*, often form patterns (Kessler, 1989; Pfennig, 1962). They are easy to see when the fluid layer is shallow, and when the illumination provides adequate contrast. These patterns are fairly regular arrays of dots or stripes whose overall diameters and spacings are usually of order millimetres (Figure 1), i.e. much greater than the size of individual cells, typically micrometres. The pattern dimensions are also much greater than the average spacing between cells ($\sim 10^{-3}$ cm). What is seen are marked fluctuations in concentration $n(\mathbf{r},t)$ of bacterial cells, as discussed in Appendix 1. Generally, high cell concentrations are associated with, and cause, downward motion of the suspending fluid. However, the velocity of the fluid, $\mathbf{u}(\mathbf{r},t)$, is correlated with both local and remote values of $n(\mathbf{r},t)$. The relations between cell concentration, mean density of the fluid and convection are also discussed in Appendix 1.

Maddock, L., Bone, Q. & Rayner, J.M.V. (ed.). *Mechanics and Physiology of Animal Swimming.*
© 1994. Cambridge University Press.

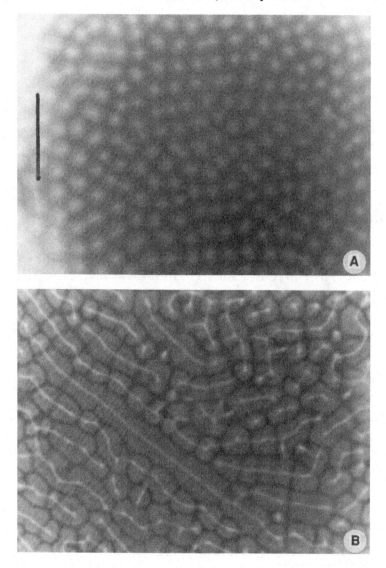

Figure 1. Bacterial concentration convection patterns seen in plan view. Dark-field illumination causes regions which contain many cells to appear white; there are fewer cells in the dark regions. (A) The fluid depth is approximately 2 mm. Scale bar: 1 cm. (B) Fluid depth 3 mm. The pattern geometry markedly depends on depth.

Physical and biological factors combine to convert an originally static microbial habitat, e.g. a quiescent fluid bacterial culture in a petri dish, into a functional dynamic system. The sequence of events leading to pattern formation is ordered, first static and then dynamic. Eventually all the ingredient phenomena occur simultaneously. Oxygen is consumed by the bacteria and supplied from the air at the surface of the culture. Thus the oxygen concentration decreases downwards from the surface. As consumption progresses the oxygen gradient steepens. The bacterial cells located within this gradient swim upwards, towards the air. Oxygen diffusion is usually too slow to reach the lower regions of the fluid. As the cells near the bottom deplete the supply, they stop swimming. The physiological components of pattern generation are respiration and oxygen-taxis. Physical factors are the location of the surface, due to gravity, and the diffusion rate of oxygen.

Initially there is no general motion of the fluid, except for the slight local fluctuations that accompany the passage of a swimming organism. As the upward swimming of the bacteria towards the air progresses, the mean density of the increasingly cell-laden fluid strata next to the surface surpasses a stability threshold. The dense fluid layer then coalesces into descending plumes. The downward motion of dense fluid is necessarily coupled with the ascent of an exactly equal volume of fluid from below. This convective motion of the fluid and its occupants, driven by them, continues for hours or days, until the culture as a whole becomes senescent.

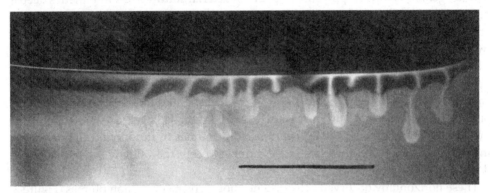

Figure 2. Concentration-convection pattern seen from the side. Air over the fluid interface. New plumes are starting while previous plumes penetrate the fluid below. The dark region between the interface and the main body of fluid is depleted of the cells which have swum to the surface. Dark-field. Scale bar: 1 cm.

Consumption and upward swimming have been recognized as 'biological'. The rest of the complex dynamics just described, the convection of fluid and the associated transport mechanisms, are entirely 'physical'. Nevertheless, there are remarkable biological consequences. There are direct benefits to the cell population in the form of an increased viability of an 'average cell'. Without convective mixing and transport, the large fraction of the cell population that is well below the surface becomes non-motile due to lack of oxygen. In contrast, in the early stages of the convection pattern, the stream carries downwards a high concentration of well-aerated bacteria, together with dissolved oxygen. This bacteria- and oxygen-transporting stream is to be seen in Figure 2 as a set of descending plumes. The displaced fluid that moves upwards is oxygen depleted and carries along some of the anoxic cells from below. Eventually, when convection is in full swing, most of the cells are again fully motile, except in the lowest strata where there is almost no movement.

EXPERIMENTAL

Two strains of motile wild-type *Bacillus subtilis* were grown in enriched minimal medium, at 37° C, in glass flasks mounted on a shaker. When the cell concentration n was sufficient, usually $n \geq 5 \times 10^8$ cells cm^{-3}, patterns formed spontaneously a few minutes after shaking ceased. Generally, patterns were observed by dark-field illumination at room temperature, in a shallow layer of fluid culture. The pattern geometry was rather depth-dependent. No patterns occurred for fluid layers shallower than ~1 mm. In thick layers the patterns could not be seen in plan view, due to excessive light scatter by the bacterial population. The time lag between first pouring the fluid culture into a diagonally illuminated petri dish and the initial observation of patterns ranged from 30 s to 5 min, apparently depending mainly on the cell concentration and the aeration of the fluid.

To understand better the relationship between upward swimming and the patterns, another geometry was employed. Normally a bacterial culture a few millimetres deep is placed in a flat-bottomed petri dish, permitting the observation of the patterns which arise in shallow fluid layers. The new geometry consisted of a cuvette constructed of two microscope slides set on edge, separated by a 1-mm thick spacer at the bottom and side edges, then sealed with silicone grease. The bacterial culture, approximately 1 cm deep, could now be observed from the side (Figure 2). It was found that, soon after the cuvette was filled, a region about 1 mm below the air/fluid interface became depleted of bacteria because they swam towards the interface. This cell accumulation at the meniscus then became gravitationally unstable. Plumes of fluid that contained many more than the overall volume-averaged number of cells then slowly began to descend downwards from the interface.

These horizontally observed cell-concentration-driven convection plumes (Figure 2) do not correspond exactly to a side view of the sort of convection rolls seen in plan view (Figure 1), because of the side wall constraint on the fluid dynamics. The thickness of the cuvette is restricted by the need for visibility and contrast. Nevertheless, it was shown that deep-layer plumes do not stir the whole fluid, whereas in shallow layers, i.e. ≤4 mm, the whole fluid eventually participates in the convective motion.

When a culture in which a pattern has previously occurred is remixed, by shaking or swirling the container, the patterns disappear. The entire fluid system becomes cloudy due to scattering of the illumination from the now uniformly distributed cells. After the container is set down, the patterns reappear. The time for reconstituting the patterns depends on the cell and oxygen concentration, the duration and intensity of the mixing, and the depth of the fluid layer. Typically that time varies between 30 and 90 s. After the patterns reappear, it takes several minutes before they are nearly steady, and often much longer before a final state is reached. Increase of the cell concentration, by growth, is clearly not a factor in pattern generation.

Water filters were used, and light intensity was kept low, to avoid thermal convection. It was also necessary to cover the observing dish so as to avoid the generation of an evaporation-induced vertical temperature gradient.

Oxygen-taxis (actually aerotaxis) of the bacteria was demonstrated using a thin layer

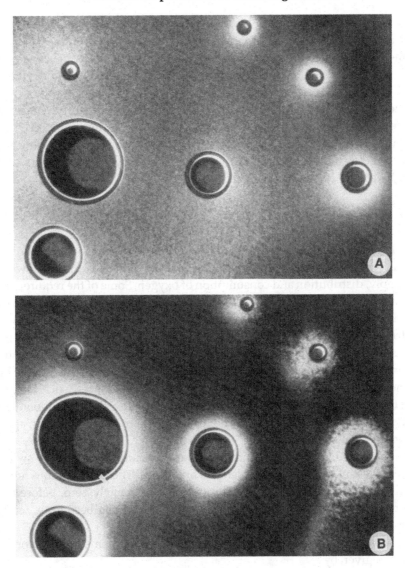

Figure 3. Oxygen-taxis of *Bacillus subtilis*. Plan view of fluid containing bacteria and air bubbles, located between glass surfaces. Dark-field illumination renders bacteria as white. The largest air bubble is ~1 mm in diameter. (A) Soon after the placement of the preparation on the microscope stage. Some bacteria have already swum towards the small peripheral bubbles. (B) Several minutes later. The spatial distribution of cells is essentially time-independent. The swarming bacteria are nearly close-packed around the bubbles.

of fluid bacterial culture between a microscope slide and a coverslip (Figure 3). Small air bubbles were trapped in the fluid. They eventually served to attract cells as the latter exhausted their oxygen supply, and oxygen diffused radially from the bubbles creating radial gradients. This test did not involve gravity or convection.

Another test of aerotaxis used the patterns themselves. It was found that when gas access to the fluid was occluded by a cover slip floating on the surface, patterns vanished underneath the central region of the coverslip. Some of the patterns from beyond the coverslip steadily persisted approximately 1 mm under the edge of it. This observation indicated that the loss of patterns below the glass cover did not imply that they were driven by surface tension gradients.

In further tests we placed the bacterial culture in a nitrogen atmosphere. Initially there were some weak and unusual patterns, presumably due to dissolved oxygen effects. Unlike in the case of an air interface, the patterns vanished after a few minutes.

Patterns generated by motility mutants differed markedly from the normal ones, but not enough data are currently available.

TRANSPORT

A complete mathematical description of consumption-driven bioconvection patterns involves six conservation equations which link the fluid dynamics, the cell's swimming, and the supply, distribution and consumption of oxygen. Some of the required ingredients are unknown, particularly a detailed description of bacterial motility in an oxygen gradient. The fluid dynamical equations provide for calculations of the fluid velocity, $\mathbf{u}$, given a driving force. For bioconvection, the driving force is gravity, operating on local departures of the fluid density from the mean. Differences of density are directly proportional to variations of cell concentration. These variations arise from the interaction of the swimming cells' trajectories with guiding influences, such as chemical concentration gradients, and with the fluid's velocity. The fluid equations are well known (Pedley & Kessler, 1992) and are not presented here. They have been applied to calculations of algal bioconvection.

The equations that govern oxygen transport and consumption, and cell swimming, can be used to estimate characteristic lengths and times. These calculations are simple in the limiting cases when the fluid velocity is approximately zero, before onset of convection, and when it is vigorous, so that complete mixing becomes a reasonable assumption.

The conservation, consumption and transport of the oxygen concentration c (molecules cm^{-3}) is given by

$$\frac{\partial c}{\partial t} = D\nabla^2 c - \nabla \cdot (\mathbf{u}c) - \gamma n. \tag{1}$$

The bacterial consumption of oxygen is γ (molecules cell^{-1} s^{-1}), the diffusion coefficient is D (cm^2 s^{-1}) and the cell concentration is n (cells cm^{-3}). If only the vertical direction is of interest, $\nabla \to \partial / \partial z$. The fluid, velocity $\mathbf{u}$, transports oxygen, hence the flux term $\mathbf{u}c$ (molecules cm^{-2} s^{-1}). For cells, the analogous equation is

$$\frac{\partial n}{\partial t} = -\nabla \cdot [\mathbf{u} + \mathbf{V}(c)]n. \tag{2}$$

The swimming velocity $\mathbf{V}(c)$ is a function of c, its derivatives, and possibly a function of the history of c in the cells' reference frame. Both deterministic and stochastic motile behaviours are included (Pedley & Kessler, 1990) in $\mathbf{V}(c)$. When $c<c$(threshold), $\mathbf{V}(c)=0$. When c is large, $V(c)\approx V_b$. In a spatial gradient of c, $\mathbf{V}(c)$ is directed towards higher c, on average.

At the beginning of a typical experiment, the cuvette or petri dish which contains the cell culture is agitated until the cells are uniformly distributed, $n=n_o$, and the oxygen concentration is $c\approx c_o$, approximately the maximum solubility in equilibrium with air. After viscous damping causes the fluid velocity $\mathbf{u}$ to vanish, eqn 1 can be used to calculate consumption times and boundary layer thicknesses.

The bulk consumption time, T, can be calculated from eqn 1 by setting $D=0$. This procedure has validity because, as will be shown, the depth of the diffusion boundary layer that develops during T is much smaller than the dimensions of the bulk fluid. This means that after the deep-lying cells consume the initial oxygen concentration they are effectively cut off from supply.

Then, without diffusion,

$$c = c_0 - \gamma\, n_o t. \tag{3}$$

When $c=c_o/2$, $t\equiv T(1/2)=c_o/2\gamma n_o$. The approximate penetration depth, i.e. the diffusion boundary layer that develops during $T(1/2)$ is $L(1/2) = [2DT(1/2)]^{1/2} = [Dc_o/\gamma n_o]^{1/2}$. Using the estimates $\gamma=10^6$ molecules cell^{-1} s^{-1} (Berg, 1983), $n_o=10^9$ cm^{-3}, $D=2\times10^{-5}$ cm^2 s^{-1}, $c_o=1\cdot5\times10^{17}$ molecules cm^{-3}, one obtains $T(1/2)\approx75$ s and $L(1/2)\approx6\times10^{-2}$ cm. Thus, by the time half the oxygen is exhausted, the new supply from the air interface has hardly begun to penetrate. The usual depth of the fluid used in the experiments is ≥2 mm, which is considerably larger than $L(1/2)$. One may calculate L for other depletions of oxygen. For example, when all the oxygen is used up, $c=0$, $T=c_o/\gamma n_o$ and $L(0)=2^{1/2}L(1/2)$, or about 1 mm. Thus, the thickness of the boundary layer, and hence the penetration of oxygen to the bulk of the fluid, is not sensitive to the assumed total consumption.

Initially, all the bacteria are uniformly distributed in the fluid. The cells located near the air interface are the first to sense the oxygen gradient. As a result, they swim upwards, towards the interface. When their concentration is sufficient, convection begins as plumes that descend from the interfacial region. The observed time for plumes to start is of order one minute, and the depth of the layer out of which bacteria swim upwards is about 1 mm, both in agreement with the calculated T and L values. The bacteria-depleted layer shows up 'black' in dark-field illumination, so that the magnitude L can easily be estimated.

The boundary gradient of oxygen can also be calculated from steady conditions, assuming that eventually there is no time dependence of c. The bacteria then consume all the available oxygen which diffuses in from the top surface. Then $\partial c/\partial t=0$ in eqn 1. At $z=0$ the oxygen concentration is approximately in equilibrium with air, $c=c_o$, and at the bottom boundary the oxygen flux vanishes, $\partial c/\partial z=0$ at $z=H$. Then

$$\frac{c}{c_o} = 1 - \frac{\gamma n}{c_o D}\left[Hz - \frac{z^2}{2} \right]. \tag{4}$$

When $c/c_o=1/2$, $z(1/2)=c_0D/2\gamma nH=L(1/2)^2/2H$; $z(1/2)=0.5\times10^{-2}$ cm for a typical value $H=3$ mm. The depth of the oxygen gradient is very shallow indeed. This calculation shows that the cell population, except for a thin surface layer, becomes anoxic in the absence of convective distribution of oxygen. The narrowness of the boundary means that it is unimportant to this calculation whether the cells and remaining oxygen in the main bulk of the fluid are stationary or well mixed, by bioconvection for example. There is, however, a big difference between the two. In the stationary case, the bacteria become anoxic and lose motility. In the convective case, the plumes and convection rolls continually pull pieces of the oxygenated boundary layer from the top surface into the bulk fluid, where they exchange places with the upward moving anoxic cell-containing streams. These regions of aerated fluid supply oxygen locally, and relatively slowly, with a characteristic time $T(1/2)$. It is this local aeration of the culture which maintains the observed motility of most of the cells in the population.

For the convective case, it can be argued that H, the depth where the flux is very small, should not be the total height of the cuvette, but some location in the moving fluid. One may set $z=H$ for some particular concentration, e.g. $c/c_o=0.1$. Then $z(0.1)=(9/5)\,L^{1/2}$, or about 1 mm, which corresponds to the size scale of some observed features of the convection.

It must be remembered that in the convective case both oxygen and bacteria are transported. Then both eqns 1 and 2 are required, with u supplied either from observations or the fluid mechanical equations. The rate of motion of the convection rolls can be experimentally determined by observing suspended particles. Preliminary measurements indicate $u\approx10^{-2}$ cm s^{-1}.

How do diffusion and convection compete over a distance R? A characteristic time for convective supply is $t_c=R/u$. For diffusion, the time is $t_d=R^2/D$. When the ratio $Pe=t_D/t_C=Ru/D$ is $\gg1$, the distribution process is dominated by convection, and conversely for $Pe\ll1$. For $u=10^{-2}$ and $D=2\times10^{-5}$, $Pe\equiv R\times10^3$. For $R>10^{-2}$ cm the long-range distribution of molecules is convective, but at small distances, e.g. in the gradient boundary layer, diffusion is dominant.

ENERGY BALANCE

The dissipation which accompanies any more or less steady convection pattern must be compensated by some source of energy. Since the system is isothermal, bacterial upward swimming, which generates gravitational potential energy, ought to be the only source of that energy. The gravitational potential energy per unit volume per second that is generated by n organisms per centimetre cubed, each organism having volume v and density, ρ_b, swimming in water with density, ρ_w is $vn(\rho_b - \rho_w)g\,V_{bu}$ (ergs s^{-1} cm^{-3}). The acceleration of gravity is g and the mean upward swimming speed of the bacteria is V_{bu}. The dissipation of energy per time per volume, due to viscosity and shear, which is the spatial variation of the fluid's velocity, $\mathbf{u}$, is $\mu(du/dx)^2$ erg s^{-1} cm^{-3}. The derivative du/dx can be estimated by dividing the observed range of u by a typical wavelength scale of the pattern, determined by measuring the distance between adjacent spots or lines. Magnitudes of u were determined from the motion of latex spheres suspended in the fluid. This motion was oscillatory, as spheres wandered back and forth in the pattern's convection rolls. A typical order of magnitude was $u=10^{-2}$ cm s^{-1}, as stated previously.

The viscous power dissipation should equal the potential energy created per time, i.e. the power generated by upward swimming. This equality actually holds at least by order of magnitude: with $u\approx10^{-12}$ cm^3, $n\approx10^9$ cm^{-3}, $\rho_b-\rho_w\approx0\cdot1$ g cm^{-3}, $V_{bu}\approx10^{-3}$ cm s^{-1}, $g=10^3$ cm s^{-2}, $\mu=10^{-2}$ g cm^{-1} s^{-1}, d$u\approx10^{-2}$ cm s^{-1} and d$x\approx10^{-1}$ cm, one obtains 10^{-4} erg s^{-1} cm^{-3} for both powers. It is actually not necessary that the power balance hold locally, since the laws of fluid mechanics convert locally exerted forces to fluid motions of much greater extent. As a matter of interest, the upward swimming power dissipation is about 10^{-13} erg s^{-1} cell^{-1}. The power required just to swim horizontally can be estimated from the Stokes drag as $6\pi\mu aV_b^2$. Taking a$\approx10^{-4}$ cm and $V_b\approx V_{bu}\approx10^{-3}$ cm s^{-1}, one obtains about 2×10^{-11} erg s^{-1} cell^{-1}. Thus upward swimming costs only about 1% more than the energy to swim altogether.

CONCLUSION AND SUMMARY

The spontaneous appearance of spatial and dynamic order, spanning distances great compared to both the size of the organisms and the mean spacing between them, suggests the existence of some form of orchestrated co-operation. Rather than that, it turns out that signalling is inadvertent, indirect and interactive with the physical environment. The primary organizational factor is the consumption of oxygen by the cell population, and its supply, unilaterally from the interface between fluid and air. The oxygen concentration gradient thus produced elicits directional motility ('aero-taxis' or 'oxygen-taxis'). As cells swim upwards and accumulate near the top surface, the mean fluid density there increases. Eventually stratification of the heavy above the light fluid becomes unstable. Plumes of heavy fluid descend, while lighter fluid from below circulates upwards. It is this convective dynamics, and the associated variations in cell concentration, that one observes.

The spatial and temporal persistence of the self-organized structure arises from the interplay of consumption, behaviour, and the long-range nature of the laws of fluid dynamics, which are coupled into the system by the force of gravity. It is especially significant that these consumption-driven dynamic structures function as an effective mechanism for improving molecular distribution and transport. The entire dynamical system becomes a respiratory engine: consumption and local depletion of oxygen elicit a complex response that supplies more of it. The overall phenomenon provides an example of the auto-constitution of a macroscopic organ, one that develops out of a cell population that organizes itself, and is organized, by consumption and the gravitational asymmetry of the physical environment.

We should like to acknowledge support from the SERC, a US National Science Foundation grant, a NASA grant, and the generous support of Ralph and Alice Sheets through the University of Arizona Foundation. J.O.K. would also like to thank the University of Leeds for its hospitality.

APPENDIX 1

Visibility of variations in concentration

Not only are fluctuations of $n(r,t)$ intrinsically of interest, they also generate contrast, since the optical mean path length is $(n\sigma)^{-1}$, where σ is the extinction or scattering cross-section. The approximate magnitude of σ is a characteristic cell dimension squared, i.e. $\sim 10^{-8}$ cm^2 for bacteria. The light intensity I, which is transmitted straight through a layer of fluid x cm deep, and neglecting multiple scattering, is

$$I(x) = I(o)\exp(-\int_0^x n(\xi)\sigma d\xi). \tag{5}$$

Equation 5 relates to the sharp contrasts which are often observed in these patterns, either with dark-field or transmitted light. There is some evidence that σ is a function of the cells' orientation. It can be affected by the local fluid motion.

Fluid density and particle concentration; sinking of concentrated regions

Why do fluid regions that contain more cells than surrounding regions sink? The mean density $<\rho>$ of fluid that contains suspended particles of volume v, concentration n and density ρ_b, where the pure fluids' density is ρ_w, is

$$<\rho> = \rho_w + nv(\rho_b - \rho_w). \tag{6}$$

As specified above, for these experimental conditions, $\rho_b - \rho_w \approx 0.1$ g cm^{-3}, $v \approx 10^{-12}$ cm^3, and $n \approx 10^9$ cells cm^{-3}. Then $<\rho> - \rho_w \approx 10^{-4}$ g cm^{-3}. Although this density offset seems small, it properly estimates the experimentally observed magnitudes of the plumes' speed of descent.

The formula used for the speed S (cm s^{-1}) was

$$S = (<\rho(n_1)> - <\rho(n_2)>)\left(\frac{gr^2}{\mu}\right) \tag{7}$$

where $n_1 - n_2 = 10^9$ cm^{-3}, and the plume radius $r = 10^{-1}$ cm. This formula provides only an estimate of magnitude. An exact formula would also depend on fluid depth, container geometry and additional factors derived from fluid dynamics. For example, 2/9 is the multiplicative factor for a solid sphere which sediments in an infinite medium.

The expressions for $<\rho>$ and S do not depend on the nature of the suspended particles, providing their number is sufficient to allow a continuum approach. It is immaterial whether the particles are passive and only undergo thermal motions or whether they swim, as do many algal and bacterial cells.

Chapter 2

Buoyancy and swimming in marine planktonic protists

C. FEBVRE-CHEVALIER AND J. FEBVRE

URA 671 CNRS, Observatoire Océanologique, 06230 Villefranche-sur-Mer, and Université de Nice-Sophia-Antipolis, 06108 Nice-cedex, France

This chapter reviews some adaptative systems of buoyancy regulation and active swimming used by marine planktonic protists to maintain position. Collective dynamics existing in experimental conditions and concentrating phenomena in natural conditions are not examined.

INTRODUCTION

Protists are predatory or photosynthetic single-celled eukaryotes (in the nanometre to millimetre range) that have physiological capabilities comparable to those in metazoans where they are distributed amongst specialized cells. Since the protists have colonized solid and fluid environments, they exhibit a wide range of behavioural patterns. To some extent pelagic species compensate for the tendency to sink by using regulatory buoyancy mechanisms. They are also capable of responding to external factors and physical parameters by adjusting their movements and swimming behaviour. The physiological activities involved in buoyancy regulation and generation of co-ordinated movements use a complex system of protein filaments, the so-called cytoskeleton, that extend throughout the cytoplasm (actin, microtubules, non-actin nanofilaments and their associated proteins) and a sophisticated machinery involving signalling molecules, cell-surface receptors, motor molecules and ions (Amos & Amos, 1991).

The present chapter reviews adaptations of individual cells and colonial forms of marine planktonic protists to buoyancy and swimming. The environmental factors

Maddock, L., Bone, Q. & Rayner, J.M.V. (ed.). *Mechanics and Physiology of Animal Swimming.*
© 1994. Cambridge University Press.

guiding the generation of concentration/convection patterns and the collective dynamics of micro-organisms that have recently been reviewed are not described here (Kessler, 1985, 1989; Levandowsky & Kaneta, 1987).

HYDRODYNAMICS OF SMALL PARTICLES

The sinking speed, v, of a single spherical particle of radius, r, and density, ρ_1, located in a fluid at rest and of density, ρ_2, is governed by the Stokes' equation:

$$v = 2\,g\,r^2(\rho_1 - \rho_2)\,/\,9\,\mu \tag{1}$$

which indicates that v is inversely proportional to the viscosity, μ, of the fluid and directly proportional to the square of the radius of the particle, to the difference in densities between the particle and fluid and to the acceleration due to gravity, g. The main forces experienced by an object moving through a fluid are the viscous forces and an inertial force resulting from acceleration of the fluid entrained by the movement of the particle. Theoretically the kinetic energy accumulated by the fluid at the front of the object would be released at the rear of the sphere. However, in real fluids such as sea-water, the kinetic energy is dissipated against viscosity. The relative importance of inertial and viscous forces varies with the Reynolds number (Re) a dimensionless quantity describing an object moving through a fluid. It is given by Re=$v. l$ / v where v is the velocity, l the length of the object and v the kinematic viscosity (viscosity of the medium divided by the density of the fluid). When Re is greater than unity, i.e. for objects such as fishes moving rapidly in water, inertial effects predominate. In contrast, if Re is much smaller than one, i.e. for small particles moving slowly through the water, such as most protists, viscous drag forces are important. A high Reynolds number indicates turbulent flow around the object while a small value denotes laminar flow. Therefore, at low Re the drag of sedimenting particles is influenced less by shape and surface configuration than at high Re (Roberts, 1981). However, when Re is close to one and size scale of the object is ≥1 mm, the influence of external morphology is presumably crucial since both inertial and viscous forces are important. Hydrodynamics at low Re have been reviewed by Lighthill (1975), Roberts (1975, 1981), Childress (1977), Levandowski & Kaneta (1987).

BUOYANCY REGULATION

Passive genomic adaptation to buoyancy

Apart from flagellates, ciliates and some other examples detailed below, most marine planktonic protists such as foraminiferans, radiolarians (polycystines, phaeodarians), acantharians and encysted forms are motionless. They have a mineral skeleton making them denser than sea-water, and so sink according to the Stokes' equation (Figure 1A,B). Because of their size, small protists (0·01-0·04 mm in diameter) such as coccolithophorids sink much more slowly than larger species (0·05-3 mm) sampled at greater depth (200-300 m), such as foraminiferans (*Hastigerina, Orbulina*), sphaerellarians

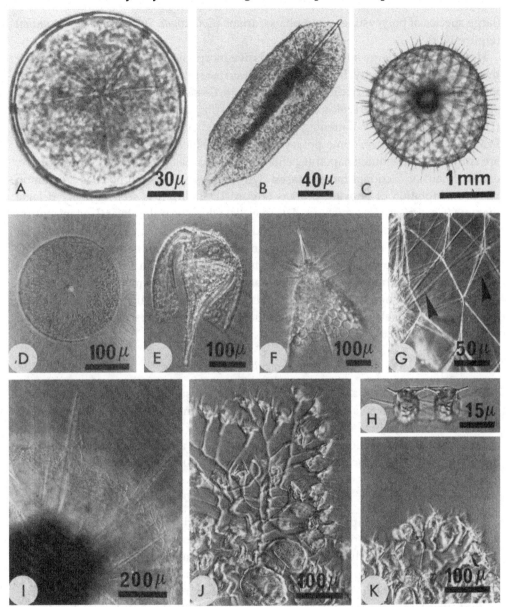

Figure 1. (A,B) Some orders of acantharians undergo gametogenesis within a cyst after profound remodelling of the cell shape. The cyst wall consists of plaques of strontium sulphate while remnants of the spicules are still visible in the cytoplasm. They sink according to the Stokes' equation. (C) In the polycystine *Aulosphaera*, the cell body is enclosed in a latticed capsule of silica. Because of their large size, they are found at great depth. (D) In the centric diatom, *Gossleriella tropica*, the thin spines that arise from the edge of the frustule increase the surface relative to the volume. (E) The dinoflagellate *Ceratium platycorne* bears flat horns that limit vertical sinking. (F) The conical shape of the polycystine Nassellaria *Pteroscenium pinnatum* may compensate for vertical sinking. (G) In *Aulosphaera* the cell body is suspended via axopodia (arrows) to the outer mineral capsule. This feature may limit vertical sinking. (H) Tiny protists, such as *Biddulphia*, sink more slowly than larger forms. (I) *Aulacantha scolymantha* is a large species which bears hollow spicules. (J,K) In the colonial peritrich ciliate *Zoothamnium pelagicum* the zooids are attached to the contractile spasmoneme (arrow). The planar organization of the colony with its rhythmic contraction-relaxation cycles and ciliary activity helps to prevent its sinking. (J) Relaxed state. (K) Contracted state.

(large species of polycystines), and phaeodarians (*Collozoum, Sphaerozoum, Aulacantha*) (Figure 1C,I).

However, many large species exhibit passive morphological features that influence their sedimenting velocity. Increase of the cell surface relative to its volume is observed in protists, especially centric diatoms (*Coscinodiscus, Gossleriella*), dinoflagellates (*Ceratium platycorne, C. praelongum, Leptodiscus, Leptophyllus*) (Figure 1D,E) and some acantharians (*Lithoptera*) which show flattening of their cell bodies (Sournia, 1986; Taylor, 1987; Febvre, 1990). Similarly, some polycystines (*Eucyrtidium, Sethophormis, Pteroscenium*) are conical or parachute shaped (Cachon & Cachon, 1986) (Figure 1F). Furthermore, enhancement of frictional forces caused by roughness of the cell body is a common passive adaptation of buoyant protists (Figure 1D,E,H,I). It is seen in many pelagic diatoms (*Gossleriella, Thalassiothrix, Asterionnella*) which bear thin lateral extensions on the edges of the frustules (Figure 1D) and dinoflagellates which exhibit horns and spines (*Cladopyxis, Ceratocorys*). Similarly, in all species of acantharians, phaeodarians and most polycystines the skeleton consists of spicules and/or cages bearing lateral teeth, extensions or ramifications that increase viscous drag and may partially compensate for the weight of the skeleton. Moreover, in some genera (*Aulosphaera, Aulacantha, Sticholonche*), the skeleton is hollow, to reduce its weight (Figures 1C,G,I & 2F (Cachon & Cachon, 1984, 1986).

Physiological regulation of buoyancy

Diverse physiological adaptations compensate for sinking in planktonic protists. They consist either of an accumulation of low density metabolic products, a release of substances serving as ballast; or an active remodelling of cell shape and cell volume. As in other protists, intracellular transport accompanying all these activities needs to be driven by the cytoskeleton and associated motor molecules (Schliwa *et al.*, 1991).

Accumulation of substances modifying cell volume and cell density

The accumulation of lipid droplets (often pigmented) provides some static lift for such protists as acantharians, polycystines and phaeodarians which have a dense mineral skeleton. Similarly, secretion of substances of lower density than sea-water is sometimes used by buoyant protists. In *Hastigerina* (Foraminifera), *Thalassicola, Thalassophysa*, and *Thalassolampe* (Phaeodaria) secretory vesicles accumulate in the cortical cytoplasm forming a thick layer of large bubbles that are continuously released at the periphery of the layer or periodically abandoned. Removing this layer (by pipetting) increases the velocity of sinking (Bé & Anderson, 1976; Anderson, 1983; Cachon & Cachon, 1984). The vesicles are transported centrifugally along a system of thin ramified or unramified hair-like processes and cytoplasmic networks (filopodia, axopodia, reticulopodial networks). Microtubule arrays are the major cytoskeletal components of these processes, and support the translocation of secretion vesicles.

In the benthic multi-chambered foraminiferans *Tretomphalus* (Myers, 1943) and *Rosalina* (Sliter, 1965), gas (probably carbon dioxide from respiration Anderson, 1987) is secreted

Figure 2. (A-C) The acantharian *Stauracon pallidus* shows rosettes of contractile myonemes (arrows) attached to a thin outer pellicle, the periplasmic cortex and to the spicules. (B) Relaxed myonemes; (C) contracted myonemes. (D) In tintinnid ciliates such as *Cyttarocylis brandti*, the cell is limited by a capsule or lorica. The oral membranelles (arrows) are used for food capture and active locomotion. (E) In *Dunaliella* sp. the two flagella (arrows) execute latero-posterior beats that push water posteriorly. (F) *Sticholonche zanclea* uses its axopodia (arrows) as oars to propel the cell through water. (G-I) Transformation of the acantharian *Heteracon mülleri* into the crawling form litholophus (H) and the precystic form (I).

just before gametogenesis in a closed chamber surrounded by a new ventral chamber. This allows the gamete-forming stage to be carried up to the water surface.

Diatoms change their buoyancy by slowly adjusting the ionic concentration of their vacuoles (Round & Crawford, 1990). Similarly, some dinoflagellate families (Noctilucidae) have adapted to planktonic life by differentiating a prominent system of turgid vacuoles, endoplasmic reticulum, and mitochondria embedded in a substance rich in monovalent cations of lower density than that of sea-water (Kesseler, 1966). Other species of marine dinoflagellates have one or two pulsatile vacuoles named 'pusules'. Pulsation of the pusules is irregular, caused by contraction-relaxation cycles of a striated rootlet of non-actin filaments. Although pusules are mainly involved in excretion, they are also implicated in osmo- and hydrostatic regulation (Cachon *et al.*, 1970, 1983).

The superficial cytoplasm itself may perform a hydrostatic function, especially in species where axopodia and filopodia are very numerous. Such is the case for some acantharians (*Acanthochiasma*, *Astrolonche*) and most radiolarian species that exhibit very dense arrays of contractile axopodia radiating from the cell body in quiescent surroundings. Upon extension, the volume of the protist is increased two- or three-fold. In the phaeodarian genus *Aulosphaera* for instance, the cell body is suspended by axopodial strands within a large spherical cage (0·5-3 mm in diameter) consisting of delicate hollow pieces of silica that delineate a hexagonal lattice (Figure 1C,G). When fully extended, almost the whole cage is filled with axopodia, while on retraction the cell volume is reduced, suggesting that sinking velocity may be slower before retraction than after. Similar adaptation is seen in pelagic foraminiferans where filopodia pass across the aperture and the pores of the shell and extend up to millimetres around the shell (Anderson, 1987).

Some amoebae (*Vexillifera*, *Mayorella*, *Boveella*) are capable of reversibly changing their shape and motile behaviour upon mechanical or chemical stimulation. Unstimulated cells move along the bottom using rounded processes (lobopods), while stimulation triggers the formation of long conical or pointed microspikes. This reaction is accompanied by detachment of the specimens which then float. Recovery of the amoeboid shape and return to their initial situation seems to take longer than the benthic to planktonic transformation, suggesting some mechanisms of buoyancy regulation.

Cell contraction inducing change of the cell shape and volume

Active buoyancy regulation takes place in diverse protists, such as dinoflagellates, acantharians and ciliates, which use original systems of contractile organelles (myonemes, spasmonemes, striated flagellar roots, subcortical fibrillar layers). Although these structures are not all similar, they share common physiological and biochemical characteristics: they consist of filaments that are thinner than actin (2-6 nm in diameter), highly resistant to denaturing agents, and are calcium-sensitive. Rapid shortening is mediated by supercoiling and twisting of the filaments (Amos, 1975; Febvre, 1981; Febvre & Febvre-Chevalier, 1982, 1989a,b; Salisbury & Floyd, 1978; Salisbury, 1983). The biochemical characteristics are known for spasmonemes (spasmin) (Amos *et al.*, 1976), flagellar roots (centrin) (Salisbury et al., 1984, Salisbury, 1989) and the retractile fibres of some ciliates (spasmin-like P23 protein) (David et al., 1992). The protein composition is rather simple (molecular weight 20,000-23,000). Spasmoneme extension does not require ATP for continued cycles of contraction and extension, whereas flagellar roots do.

In most dinoflagellates, the cell body is limited by a complex cortex that includes a layer of rigid plaques of cellulose, a sheet of microtubules and a dense layer of filaments. The plaques form a rigid theca which prevents change of cell shape (Dodge & Greuet, 1987) (Figure 1E). However, in the Gymnodinides, the rigid plaques are lacking and myoneme bundles are attached to the inner sides of the cell membrane. Contraction and relaxation of the myonemes is capable of changing the shape and volume of the cell body (Cachon et al., 1987). They also cause a cylindrical tentacule to contract in *Noctiluca* (Soyer, 1970) or move a large lobe slowly in *Pomatodinium* (Cachon & Cachon, 1985). Although contraction-extension appears mainly implicated in feeding it may also play a part in buoyancy or participate in swimming concurrently with the flagella (Cachon & Cachon, 1986). In Leptodiscinae species, rhythmic and powerful contractions propel the cell body by jerky movements like jellyfish (Cachon et al., 1987).

In acantharians inflation-deflation of the cell surface is performed via contraction and relaxation of 40-1000 large contractile myonemes that suspend a superficial thin elastic pellicle (periplasmic cortex) to the tip of 20 radial spicules (Febvre, 1981) (Figure 2A). The myonemes participate in two types of contractions: slow undulating movements (1-10 min duration) and sudden contraction (10 ms range). The undulating movement consists of stepwise contractions and partial relaxations leading the myoneme to become progressively taut between its anchorage points. During this movement, the position of the periplasmic cortex and the cell volume remain unchanged. The myonemes show an alternation of clear and dense bands whose spacing varies with the amplitude of contraction. They consist of thin nonactin filaments (2-4 nm in diameter) twisted in microstrands (Febvre & Febvre-Chevalier, 1982; Febvre-Chevalier & Febvre, 1986). During the slow contraction-relaxation, coiling interaction of the twisted microstrands occur at the level of the myoneme filaments and as revealed in polarized light, the molecular orientation of the filaments is modified (Febvre et al., 1990). Such a structural change depends only on the binding of internal calcium ions to myoneme filaments without the requirement of ATP (Febvre & Febvre-Chevalier, 1989a). Mechanical stimulation induces Ca^{2+}-influx across the cell membrane eliciting rapid synchronous contraction of all the myonemes (Figure 2B,C). The cortex is thus pulled rapidly towards the apex of the spicules which induces a rapid inflation of the protist (Febvre & Febvre-Chevalier, 1989b). This change of cell volume may be involved partly in excretion or release of defaecation vacuoles and in buoyancy regulation.

Peritrich ciliates exhibit rapid contraction-relaxation cycles of the cell body implicated in buoyancy regulation. This behaviour is exemplified by the planktonic colony *Zoothamnium pelagicum* which consists of a ramified contractile stem to which vegetative and reproductive zooids are attached (Figure 1J,K). As in the benthic peritrich genera *Zoothamnium*, *Carchesium* and *Vorticella*, periodic contraction-extension is brought about by coiling or folding of a large bundle of spasmin filaments (spasmoneme) (Amos et al., 1976) (Figure 1J arrow). This rhythmic activity produces a jerky movement of the colony which changes alternately from spherical to parachute-shaped. The metachronal beating of the oral ciliature also plays an active part in transporting the colony through the water. Furthermore, the adaptation of *Zoothamnium* to planktonic life is improved by the planar organization of the colony which floats facing the current (Laval, 1968).

SWIMMING BEHAVIOUR

Unlike buoyant protists that minimize sinking by increasing frictional interactions with the fluid, motile cells minimize drag by developing hydrodynamic profiles. For instance, large ciliates appear to be close to the optimum shape for the appropriate Reynolds number and display drag 10% lower than that of equivalent spherical protists (Roberts, 1981). Three types of movements can be used by protists for propulsion through the water: (1) rhythmic beating of appendages such as cilia and flagella; (2) rowing motion of rigid axopods used as oars; (3) crawling locomotion.

Ciliary and flagellar beating

Cilia and flagella are long, cylindrical, hair-like or whip-like appendages (0·2 μm in diameter; 10-100 μm long) that project from the cell surface of eukaryote cells such as ciliates and flagellates. These appendages generate currents around the cell which are used for compensating sinking and for other functions such as feeding, cleaning, avoidance reactions or escape movements. In the last two decades, attempts to create theoretical models which reproduce the observed oscillatory beating of ciliated and flagellated cells have been made by Machin (1958), Lighthill (1975), Roberts (1975, 1981), Childress (1981). Structural and molecular data have provided better understanding of the functioning of cilia and flagella (Gibbons, 1989; Witman, 1990), providing new data or understanding of the influence of physical requirements for bend propagation (Dentler, 1987; Holwill, 1989; Sleigh, 1989a,b).

The bending motion of cilia and flagella results from sliding of 9 adjacent doublets of microtubules via dynein side-arms (axoneme) (Goodenough, 1989; Satir, 1989). Dynein is a large protein complex consisting of 2 or 3 globular heads, linked by flexible strands (Goodenough & Heuser, 1989). The bending force is produced through a regular cycle of conformational change driven by ATP-binding and dephosphorylation. A central pair of singlet microtubules surrounded by a sheath rotates during the beat. In addition, radial spokes, microtubule-microtubule linkers and microtubule-membrane filaments (some with elastic properties) influence the propagation of the bending waves.

Because cilia are of equal size, shape and structure, their movement is remarkably homogeneous among ciliated cells. It consists of a cyclic gyration, directed counterclockwise when observed from tip to base, and can be divided into two phases: (1) an effective stroke during which the cilium is first straight, then curves at its base, taking the shape of an S; (2) a passive recovery stroke during which a bend is propagated along the appendage. This recovery stroke is slower than the effective stroke. When the ciliate swims forwards, the active stroke is oriented backwards. Backward swimming is caused by reversal of ciliary beating and forward orientation of the active stroke (Naitoh, 1982). During the effective stroke, all the doublets in the upper half of the axoneme (doublets 1-5) are synchronously activated, while during the recovery stroke the other doublets are activated. For individual cilia and flagella the value of the Reynolds number is low (10^{-5}-10^{-6}) (Sleigh, 1984). The active momentum consists of two elements, one counteracting the passive elastic forces, the other straightening against the viscosity of the fluid. The maximum elastic momentum develops at the apex of the waves, while the maximum viscous momentum of the curvature takes place half way between two successive crests.

The axoneme of cilia or flagella is linked to a permanent feature, the basal body. In the ciliates, the basal bodies are connected by a complex system consisting of overlapping microtubules and/or strands of non-actin filaments that underlie the cell membrane (Grain, 1987). The distribution of both cilia and the associated subcortical system (kineties) controls the swimming behaviour. According to the spacing of cilia in the kineties, the movement of one cilium interferes with that of the adjacent ones, modifying the drag on the water (Sleigh, 1984, 1989a,b). This can lead either to the formation of a synchronized and powerful propulsive movement or to a weaker delayed propulsive force giving co-ordinated metachronal waves (Machemer, 1974). In the oral region of many ciliates (heterotrichs, peritrichs, oligotrichs, tintinnids), the cilia are closely associated in rows of membranelles whose beating drives food particles towards the cytostome and generates a helical forward path (Figure 2D, arrows).

Ciliates are highly sensitive cells, capable of rapidly adapting their path to local environmental change. Experimental analysis of swimming behaviour and vertical distribution of planktonic ciliates indicates that the average upward velocity of oligotrich ciliates such as *Laboea* and tintinnid ciliates such as *Parafavella* or *Tintinnopsis* is very rapid, ranging 1-2·5 m h^{-1} (Johnson, 1989). The response to mechanical stimulation has been extensively studied in benthic or sub-benthic species such as *Paramecium, Stylonichya, Euplotes* and *Tetrahymena* (for reviews, see Naitoh & Sugino, 1984; Machemer, 1988; Deitmer, 1989). Depending on the nature and duration of the stimulus, two kinds of responses have been described: change in swimming velocity, and avoidance reactions. Change in swimming velocity takes place in response to anterior or posterior stimulation of the protist. Active forward swimming is correlated with posterior stimulation that triggers hyperpolarization of the cell membrane. Reverse slow backward swimming is associated with anterior stimulation and membrane depolarization. The avoiding reaction is induced by stimulation to the anterior end of the ciliate. It triggers ciliary reversal and backward excursion, followed by re-orientation and normal forward swimming. This reaction can be reproduced by a depolarizing current pulse which opens voltage-dependent calcium-channels. The calcium current is exclusively localized at the base of the ciliary membrane (Dunlap, 1977). It raises internal concentration of cytosolic calcium ions above 10^{-6}M. The direction of beating (forward or backward propagation waves) and arrest are controlled by activation of individual dynein arms under regulation of intracellular calcium ion concentration (Stephens & Stommel, 1989). Beat frequency appears to be controlled by cyclic AMP and cyclic GMP and influenced by other chemicals such as folic acid. These substances act as attractants,

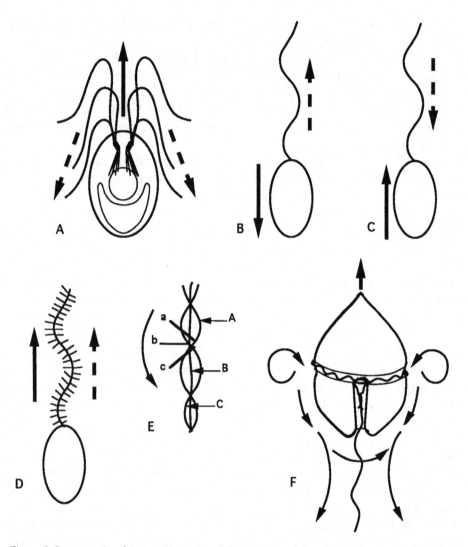

Figure 3. Interpretative drawing illustrating different types of flagellar movements. Full arrows, direction of propulsion. Dotted arrow, propagation of the wave of flagellar beating. (A) *Chlamydomonas* type; (B) pulsella; (C) tractella; (D) flagellate with a flagellum bearing mastigonemes. (E) During the three positions A, B, C of the flagellum, the mastigonemes take the positions a, b, c. The arrow indicates their displacement. (F) Dinoflagellate seen in ventral view. The curved arrow indicates the direction of rotation.

probably signifying the presence of food in the fluid environment, and modify swimming speed. However, since deciliated ciliates are still reactive, the ciliary membrane may not intervene directly in chemoreception (van Houten, 1989).

Flagellates can show important variations in locomotory behaviour and in the Reynolds number at which they operate (Jahn & Bovee, 1967; Levandowsky & Kaneta, 1987). Depending on their length, position, number and on presence or lack of additional structures, the flagella can propagate movements like typical cilia or execute undulatory waves that push or pull the cell in one direction. They can also induce spiral swimming patterns. In biflagellated protists such as *Platymonas*, *Chlamydomonas* or *Dunaliella* the flagella execute latero-posterior beats that push water posteriorly (Figures 2E & 3A). Because of the low Reynolds number, the resulting locomotion is a jerky breast-stroke with mean speed up to 200 μm s^{-1}. In these flagellates, rhizoplasts (descending fibres or nucleus-basal body connectors) extend from the basal bodies through the cytoplasm. They consist of centrin-based striated rootlets (Salisbury *et al.*, 1984). In some species such as *Platymonas subcordiformis* they influence *in vivo* the pattern of co-ordinated movements (Salisbury & Floyd, 1978). In *Chlamydomonas* centrin-based fibres are responsible for nuclear movement but may not induce shape-change (Salisbury, 1989). In many flagellates a distal connecting striated fibre links the two adjacent basal bodies to one another. It was recently shown in *Spermatozopsis* that contraction of this centrin-base fibre mediates large changes in flagellar orientation which take place during photophobic responses *in vivo* (McFadden *et al.*, 1987). However, because of the presence of a rigid cell wall there is no flagellar reorientation in *Chlamydomonas*.

Flagella that push the cell through water are called pulsella. Pushing takes place when forces are exerted towards the base of the flagellum which drives water away (Figure 3B).

Flagella that pull the cell through water are called tractella. They are oriented ahead and the helical waves propagate from the tip to the base of the appendage (Figure 3C). This motion produces a tractile force directed away from the base. Flagella bearing at the surface of their cell membrane a cover of stiff processes or mastigonemes perform waves, directed distally, which drag the body through forces pushing the water towards it (Figure 3D,E). These features may provide resistance to shear and produce more complex patterns of beating of the flagella.

Such a swimming motion is illustrated by the helioflagellate *Ciliophrys marina*. This species performs reversible transformations, passing rapidly from a spherical heliozoan form with axopodia and a single nonmotile flagellum in the form of a figure-8, to a flagellate form. The heliozoan form is observed when sea-water is undisturbed, while shaking triggers unrolling of the flagellum. The pulling forces are exerted on the body by an anteriorly-held flagellum bearing a double row of tubular mastigonemes (Davidson, 1982). Planar flagellar waves propagate from the base to the tip. They pull the cell body forwards very rapidly (up to 150 μm s^{-1}) in a slightly curved path, while water flows towards the cell. During this movement the axopodia become trailing, then retract while the cell body elongates. The avoidance reaction occurs when the cell is stimulated. During this event, the flagellum retracts into the figure-of-eight, then re-extends after a delay. Several cycles of contraction-extension of the flagellum can be performed to re-orient the cell.

A typical helical swimming pattern characterizes most dinoflagellates that normally swim rapidly. The cell body in these protists is generally top-shaped, divided into two halves by a transverse groove around the body (Figure 1E & 3F). One of the two flagella is transverse and ribbon-shaped and lies in the groove. It includes a typical 9+2 axoneme on the outer edge and a striated fibrillar strand lying on the inner edge. The strand is shorter than the axoneme, which is always folded within the groove and undulates continuously from the base to the tip. Since removal of the other flagellum does not stop the cell, helical waves along this transverse flagellum are mainly responsible for a rotary forward locomotion. The other flagellum is longitudinal, emerging normally to the transverse one. It projects in general posteriorly and is supported by a typical axial core of microtubules and associated structures. The longitudinal flagellum propagates sinusoidal waves which are three-dimensional. In *Ceratium* the longitudinal flagellum contains a large contractile paraxial fibre which pulls the appendage into a flagellar pocket close to the cytostome, while it does not in *Oxyrrhis marina* (Maruyama, 1981, 1982, 1985; Cachon *et al.*, 1988; Cosson *et al.*, 1988) This movement may be involved in food capture rather than in swimming (Cachon *et al.*, 1991).

Rowing movement of axopodia

The marine pelagic heliozoan *Sticholonche zanclea* has a remarkable swimming system (Hollande *et al.*, 1968; Cachon *et al.*, 1977). The cell body is oblong, measuring around 200 µm. It bears long, stiff axopodia and groups of spatula-shaped spicula (Figure 2F). Each axopod is stiffened by a large bundle of relatively stable microtubules inserted in a dense basal material which articulates into cup-like depressions on the surface of the nucleus (Figures 2F & 4). Bundles of thin non-actin filaments (2-3 nm in diameter) connect the base of the axopodia to the periphery of the depressions. These filament bundles are able to coil, spiral or fold during inclination of the base in the caveola and their maximum shortening ratio is one-tenth of their extended length. Bundles of external filaments connect the axopodia to one another so that the axopodia are capable of performing co-ordinate movements The axopodia in *Sticholonche* are used as oars for propelling this protozoan through the sea-water. The axopodia are arranged in two groups, a non-motile dorsal group, and a motile lateral group (latero-dorsal, lateral and ventral). Metachronous beating (1-3 strokes per second) starts from the anterior to the posterior ends of the protist. The motile cycle consists of two phases, an active stroke (0·04 s), and a recovery stroke (0·2-0·4 s) separated by transitory pauses. During the active stroke *Sticholonche* moves forwards, then stops; during the recovery stroke it retreats slightly, and pauses before starting a new locomotory cycle. The resultant movement is rather an inefficient forward propagation.

Crawling motion through water

Crawling motion through water is illustrated in some species of acantharians such as *Heteracon mulleri, Conacon foliaceus, Stauracon pallidus*. In these species the vegetative form is spherical (Figure 2A,G) and the cell body is sustained by 20 mineral spicules embedded at their base in a fibrillar material. Just before gametogenesis, the spicules re-orient slowly so that the acantharian resembles a closed umbrella (Figure 2H,I) (Febvre, 1990). This is especially obvious in the species *Heteracon biformis*, which bears 19 thin, short spicula, and one which is spatula-shaped and much longer than the others. The cell body (60-70 µm in diameter) becomes oblong, lengthened in the direction of the longest spicule. The outer structures (myonemes and periplasmic cortex) are dragged

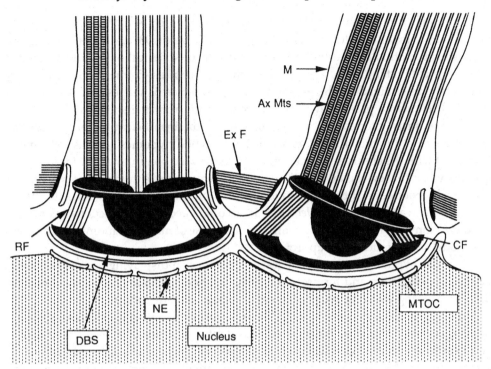

Figure 4. Interpretative drawing illustration the structure and the beating movement in the heliozoan *Sticholonche zanclea* (according to Cachon *et al.*, 1977) (see text). N, nucleus; NE, nuclear envelope; DBS, dense basal substance; MTOC, microtubule organizing centre; M, cell membrane; Ax Mts, axopodial microtubules; Ex F, external filaments; CF, contracted nonactin filaments; RF, relaxed nonactin filaments.

posteriorly. The axopodia, which arise radially in the spherical form, follow the same path, so that the cell centre turns towards the anterior. This new form called 'litholophus' has become much smaller than the vegetative stage and starts crawling forwards very slowly (1-5 µm s^{-1}) along a straight path through the water. This forward locomotion can stop and attempts to return to the centric shape can occur twice or more, then the protist stabilizes definitively in the litholophus form. The forward movement of litholophus is smooth, accompanied by centrifugal transport of very thin particles along the trailing axopodia. Exocytosis, translocation of matter, and shedding of the superficial structures (myonemes and cortex) last 10-20 min. Even the short spicula are released backwards one by one, while thin, oval, birefringent plaques are formed in Golgi vesicles, then progressively arranged at the cell surface, announcing encystment. The Reynolds number being probably very low, the cell is at rest if there is no driving force. This is confirmed by the lack of forward motion when litholophus is dead. This suggests that the forward motion of this cell may not be purely passive. Reversible cell-shape change from the spherical to the oval outlines, and conspicuous backward flow of cellular material taking place during the transformation, may produce sufficient thrust to generate the very slow forward motion observed in *in vivo* preparations.

Two kinds of physical models have recently been proposed for explaining self-propulsion at low Reynolds number. Shapere & Wilczek (1989a,b) calculated the net translational motion caused by small deformations of spherical or cylindrical bodies and suggested that in "the absence of inertia the motion of a swimmer through a fluid is completely determined by the geometry of the sequence of shapes that it assumes". Although *Heteracon* passes reversibly twice or more from the spherical to the oval shapes, the thrust exerted on the cell surface may be insufficient for propelling the protist because this transition is very slow. The second hypothesis seems to be more consistent with the observations. It suggests that directed membrane flow (as involved in endocytosis/exocytosis or osmotic regulation) may induce a fluid motion outside the cell which produces a thrust for motion exerted on the cell surface (Sornette, 1989). Since the forward motion in *Heteracon* is accompanied by conspicuous exocytosis (visualization with Indian ink as a marker) and release of matter including the spicules, this model seems much more appropriate.

A similar hypothesis may be suggested for amoebae displaying a transition from benthic to pelagic life and for protists bearing long cytoplasmic processes (all pelagic Sarcodina) along which active permanent transport of matter occurs.

Although protists are generally unable to perform efficient and long-lasting swimming against currents, they have developed adaptive systems of buoyancy regulation and bear locomotory organelles used to compensate for sinking. In most swimming protists the cell membrane is excitable and sensitive to diverse stimuli. Membrane potential change elicits receptor or action potentials that trigger rapid changes in the swimming activity, and modifies both direction and velocity of the path. Many flagellates have special light receptors that are generally closely associated with the effectors (flagella). Studies of the dynamics of individual trajectories has allowed understanding of their directed locomotion and gave useful indication of their collective behaviours (Kessler, 1985). In natural conditions, the most evident behavioural response is generated by the daily regular alternation of light and dark which induces diurnal migration of planktonic protists. Although the cilium and flagellum are by far the most ubiquitous, efficient and well-documented apparatuses, other less common swimming systems have been described in the present chapter to illustrate the remarkable adaptation of protists to planktonic life.

Chapter 3

The role of fins in the competition between squid and fish

J.A. HOAR, E. SIM, D.M. WEBBER AND R.K. O'DOR

Biology Department and Aquatron Laboratory, Dalhousie
University, Halifax, Nova Scotia, B3H 4J1, Canada

In recent decades cephalopod locomotion studies have focused on jet propulsion, largely ignoring the lateral fins used by almost all squid during swimming. There are many types of fins, such as large rhomboidal fins, smaller triangular fins and long marginal fins. Not only does fin shape vary between squid species, but also fin usage varies, often with the lifestyle of the squid. The shape of the fins also changes during ontogenetic growth, and nearly all hatchling squid have fins proportionally smaller than those of adults. This change in relative size may also reflect a change in the use of the fins similar to the different usages of the variously shaped adult fins. Understanding the significance of fin design will improve our ability to predict the lifestyles of rarely seen squids. Since fins are used in synchrony with the jet, there is the potential for both synergistic and antagonistic modes of use. Knowledge of the relative importance of fin thrust in squid swimming is also essential for interpreting the new data on jet pressures available from field tracking studies.

INTRODUCTION

For hundreds of millions of years cephalopods roamed the world's oceans as jet-propelled masters of the pelagic world, until fishes, using highly efficient undulatory locomotion, ousted them from many preferred nektonic niches. These two groups have been the focus of comparative reviews (Packard, 1966, 1972; O'Dor & Webber, 1986) which integrate the lifestyles of cephalopods with their ecology, anatomy, physiology and locomotion. There is a vast literature on the swimming of the fish interlopers (Blake, 1983a), from typical body undulations (Magnuson, 1978; Webb,

Maddock, L., Bone, Q. & Rayner, J.M.V. (ed.). *Mechanics and Physiology of Animal Swimming.*
© 1994. Cambridge University Press.

1978c, 1988) to more specialized flapping fins (Daniel, 1988) and undulatory marginal fins (Blake, 1983b; Lighthill & Blake, 1990); this includes both theoretical and direct measurements of energetic costs and efficiency. Quantitative studies of cephalopod locomotion started later and have been focused on jet propulsion (Trueman & Packard, 1968; Johnson et al., 1972). In the last decade, methods for telemetering jet pressure from freely swimming cephalopods, combined with video analysis and swim-tunnel respirometry, have improved our knowledge of their energetics (Webber & O'Dor, 1985, 1986; O'Dor, 1988b; O'Dor et al., 1990; O'Dor & Webber, 1991). More recently, telemetry studies have been used to monitor lifestyles and energetic costs in the oceans (O'Dor et al., 1993a,b), providing insights into the relative frequency of occurrence of various types of locomotion.

The study of locomotion in cephalopods is important because of their remarkable versatility as swimmers, with the ability to change direction easily, rise or dive almost vertically, hover in one spot, pounce sideways abruptly, stop instantly and even reverse. These talents aided the early cephalopods in hunting and escaping from predators. Information on cephalopod locomotion, lifestyles and energy budgets in the wild can be used to understand natural population dynamics and ecology as well as the effects of physical and biological interactions on the evolution of all nektonic animals. Neutrally buoyant Nautilus appears to be more economical than fishes at low speeds (O'Dor et al., 1990), but performance data on negatively buoyant squids suggests that, at higher speeds, they lose in direct competition with fishes because of the inherently low Froude efficiencies of jet propulsion (O'Dor & Webber, 1991). Despite remarkable adaptations of their molluscan heritage for power production (O'Dor, 1988a), squids cannot match undulatory fishes for speed or economy of movement. The jet does, however, provide acceleration and manoeuvrability better than those of generalist fishes (Foyle & O'Dor, 1988) and, perhaps, better than fishes in general. Wells (1990) notes that although evolving cephalopods have developed a variety of other means of locomotion, none has abandoned the jet, which has unique advantages for escape and respiration.

Recently, better descriptions of the fins as muscular-hydrostats (Kier, 1989; Kier et al., 1989) and new models of undulatory fin propulsion (Daniel, 1988; Lighthill & Blake, 1990) have become available, which should make better analyses and comparisons of squid and fish possible. Most cephalopods have fins which can undulate, but none uses them as an exclusive mode of locomotion. This suggests either that cephalopod undulatory fins are less effective or less efficient than those of fishes or that the combination of fin and jet is more effective or efficient than fins alone. If the jet functioned only for escape, it is hard to imagine its value for a squid like Thysanoteuthis rhombus which has so much fin that it can barely move by jet. This chapter examines some of the circumstances in which the jet and fins act synergistically and relates swimming to behaviour. Our interest in reconstructing squid behaviour and energy consumption patterns from telemetered mantle cavity pressure records requires that we understand whether jet and fin activities are integrated or independent.

WHAT IS WRONG WITH JETS?

Evolution of cephalopods is thought to have begun half-a-billion years ago when animals expelled water from their shells as they pulled their head and body in to avoid predators. Although it was probably highly unpredictable motion with little or no control, the ability to move above the sea floor to escape from benthic predators or to be able to see prey from above must have made the trick worthwhile. Along with the evolution of near-neutral buoyancy through the development of a chambered shell, this ability would have given these early cephalopods a distinct competitive advantage and led to the evolution of a wide variety of forms, including the highest animals in the sea at that time (Packard, 1972). These cephalopods dominated the rich upper layers of the oceans, due to the freedom of movement that they enjoyed, until the advent of fish with highly efficient body / fin undulatory movements forced cephalopods out of their niches and into the more peripheral environments in which we find them today (Denton, 1974).

Early cephalopods, although small, possessed relatively large external shells, similar to modern nautiloids; this limited jet propulsion, due to the small space for the mantle cavity and the large drag and inertial forces resisting movement of the shell (Denton, 1974). It is not known how ammonites swam, but they may have evolved further down the squid path than is generally thought by internalizing shells and externalizing mantles (Jacobs & Landman, 1993). By giving up the external shell, coleoid cephalopods released constraints on their locomotor systems and have been able to develop powerful propulsive muscles (Bone et al., 1981), large mantle cavities and alternative buoyancy systems. Zuev (1965b) realized that these animals, too, were caught up in an anatomical paradox, with the need for large water reservoirs for jetting vying with the need for more muscle to attain greater speeds. Since most squids have mantle cavities that are almost half of their total body volume, this is a serious displacement of muscle bulk. In comparison, fish with the same frontal area and drag do not have to waste internal space on a pressurized water reservoir and are nearly solid muscle in cross-section, giving higher performance and efficiency.

There are many planktonic and mesopelagic cephalopods, such as the cranchids and probably Architeuthis, which store ammoniacal fluids of lower density than sea water in reservoirs or tissues to give them neutral buoyancy (Clarke et al., 1979). This also displaces either mantle volume (cranchids; Clarke, 1966) or muscle mass (Architeuthis; Roper & Boss, 1982), causing another trade-off (Zuev, 1966). Alexander (1966) quantitatively explored the trade-offs of increasing drag versus dynamic lift for fishes using gas bladders and lipid stores, but there is no such analysis for squids. Active secretion of gas into soft-walled bladders would probably prove very expensive for vertically migrating squids, but the directionality of the jet combined with lift production by the fins may make dynamic lift more efficient for squids than for fishes.

Jet propulsion is well-suited to soft-bodied animals because a deformable body allows rapid expulsion of water to produce thrust and elastic expansion for the next power stroke. Jet propulsion systems produce thrust in the opposite direction from the ejection of water, and the largest possible mass expelled with the greatest possible

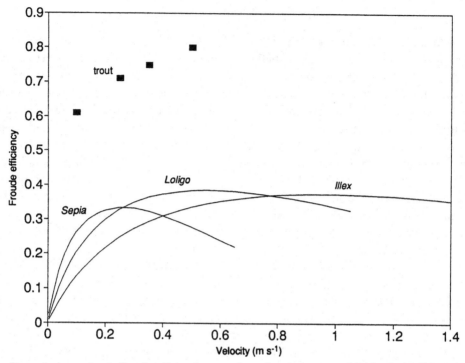

Figure 1. Approximate Froude efficiencies over the swimming ranges of 0·6-kg *Sepia officinalis*, *Loligo pealei* and *Illex illecebrosus* compared to a 0·22-kg trout (Webb, 1971). Based on jet pressure (*p*) records for animals swimming at a range of velocities (*u*) as given in O'Dor & Webber (1991). Jet velocity (m s^{-1}) is related to mantle pressure by the Bernoulli equation such that $u_j=(2p/d)^{0.5}$, where *p* (Pa) is twice the mean pressure and *d* (1025 kg m^{-3}) is the density of sea-water. Actual jet velocities (and Froude efficiencies) vary as pressure varies over each jet cycle, including zero values for over half the cycle during refilling, but this approximation is adequate for comparison (O'Dor, 1988a).

velocity will drive the animal the furthest (Johnson *et al.*, 1972; Alexander, 1977). To maximize the thrust generated during an attack or an escape, jet-propelled animals typically develop a system with separate larger inhalant and smaller exhalant apertures, which also increases efficiency (Trueman, 1975). In normal swimming, however, ejecting a mass as rapidly as possible wastes energy and is less efficient than ejecting the same mass at a velocity only slightly higher than the animal's velocity. This is defined by Froude efficiency ($E_f=u/(u+0.5u_j)$, where *u* is the animal's velocity and u_j is the jet's velocity; Alexander, 1977). Figure 1 illustrates the Froude efficiencies of different cephalopod jets at different swimming velocities. Maximum efficiencies are low compared with those of fish which may reach as high as 0·81 (Webb, 1971). In *Sepia*, funnel dimensions and mantle contraction rate produce maximum efficiency at speeds just beyond those of normal fin-driven swimming, while *Illex* does well over a wide range of higher speeds which can only be achieved by jetting.

Another important part of jet propulsion is refilling. Since no jet thrust is provided during this phase, the animal slows down and the speed fluctuates. This can be as energetically costly for the animal as drag. This idea, called the acceleration reaction, was first discussed in relation to jet propulsion by Daniel (1983) for the unsteady swimming of jetting medusae. The acceleration reaction creates a force which resists changes in the velocity of the animal (Batchelor, 1967). Daniel (1984, 1985) found, however, that the importance of the acceleration reaction diminished, the longer a medusa swam after starting from rest; although its relative importance in resisting movement is about 50% initially (the other 50% is drag). O'Dor (1988b) found the acceleration reaction during continuous swimming in squid contributed only about 5% of resistance. Thus, although the acceleration reaction dominates escape jets, it is much less important during sustained swimming of cephalopods than might be expected for a purely jet-propelled animal.

WHAT GOOD ARE FINS?

Nautilus produces two types of jet, neither of which bears much resemblance to that of the endocochleates. The powerful piston pump driven by paired head retractor muscles, which pull the head into the shell and expel the water out of the funnel (Packard *et al.*, 1980), is similar to the ancestral system used for escape. The second, highly efficient, system for respiration and slow swimming (Wells & Wells, 1985; O'Dor *et al.*, 1990) is based on adaptation of the foot into a funnel to direct water flow. Routine cruising by *Nautilus* results from modified peristaltic waves along the funnel wings which produce a continuous flow of water over the gill filaments (Wells & Wells, 1985). The funnel (or hyponome) of *Nautilus* is not a patent tube like that of the endocochleates, but consists of two flattened lobes that fold together and overlap to form a functional funnel. The low velocity waves which propel water inside the duct created by the mantle are, in fact, more like the undulations seen in coleoid fins than anything associated with jetting. If *Nautilus* could project its funnel outside the mantle, its undulations would probably provide enough thrust for swimming! This could be another option for the mysterious ammonites postulated by Jacobs & Landman (1993).

Although the shapes differ, the organization and histology of the hyponome (Hochachka *et al.*, 1978) and lateral fins of squids and cuttlefishes (Kier, 1989) are very similar. In studies of fin morphology of cuttlefish, which are easier to keep in the laboratory, Kier (1989) showed the fins to be a tightly packed three-dimensional muscular array, such that a decrease in one dimension results in an increase in another dimension. In such 'muscular hydrostats', the support necessary for movement is provided by the orthogonal arrangement of the three types of fin muscles. Kier (1989) also showed that there were connective tissue fibres present in the fin that could serve as an elastic energy storage system during fin beating, improving the efficiency of the fin musculature, much as occurs in the mantle muscle (Gosline & Shadwick, 1983). Kier *et al.* (1989) measured the activity in fin muscles and found that oxidative muscles are responsible for moderate fin movements in cuttlefish, while anaerobic glycolytic muscles power strong movements. Mechano-receptors in the cuttlefish fin oriented in three mutually perpendicular planes can detect distortion of the fin in any direction (Kier *et al.*, 1985). Since there is such fine control of fin movement, there must be

Table 1. *Cephalopod fin form and function.*

	Ommastrephid (e.g. *Illex*)	Loliginid (e.g. *Loligo*)	Cuttlefish (e.g. *Sepia*)
Fin form	Triangular*	Rhomboidal*	Bordering/Marginal
Fin waves	No	Yes	Yes
Fins flap	Yes	Yes	No
Fins wrap in escape	Yes	Yes	No
Balistiform locomotion	No	No	Yes
Clap-and-fling	Yes?	Possible	No
Flight	Yes	No?	No

*The shape of the fin varies through ontogeny, typically with fin surface area increasing faster than body length (see Figure 1 and Table 2.)

complex neural systems to co-ordinate the fin beats, but details are lacking at present, as there has been little study of fin control since Boycott's (1961) work.

Among modern cephalopods the paired lateral fins, which stabilize the jet, help to define and keep them competitive in their niche by creating special advantages and disadvantages (Table 1). The shape of the lateral fins varies in accordance with the size of the cephalopods (Figure 2) and the way of life of the animal (Packard, 1972). For instance, oceanic squids (such as the ommastrephids) have small, triangular fins that appear to function mainly as rudders, whereas the coastal loliginid squid have large, rhomboidal fins useful for propulsion or 'soaring' in currents (O'Dor et al., 1991, 1993b). The cuttlefish's undulatory marginal fins, with their very complex wave-producing ability, provide efficiency and manoeuvrability at low speed for the animal's coastal, bottom-dwelling lifestyle. There is no single optimum fin size or shape. Large loliginid fins are more useful at the lower speeds of a coastal environment, but also produce more drag at higher speeds. The small fins of the oceanic squids produce less drag at high speed, but force these animals to depend more on the inefficient jet-propulsion system.

Undulatory fin locomotion is more efficient than jet propulsion because forward momentum is produced at all times, unlike the unsteady jet cycle. Cephalopod lateral fins not only enhance stability during movement but also produce undulatory locomotion. Beyond noting that they are rolled against the mantle at high speeds (Webber, 1985; O'Dor, 1982; Bradbury & Aldrich, 1969; Zuev, 1965a), little quantitative work has been done to indicate what fins do best. After amputation of the fins, *Illex illecebrosus* (Williamson, 1965), *Symplectoteuthis oulaniensis* and *Illex coindetii* (Zuev, 1966) were incapable of translational movement. Although they could attain similar speeds to normal squids, they tended to follow a sinusoidal curve, rather than a linear path, and gradually sank. This indicates that the fins are important, not only for steering, but also for generating lift. Zuev (1966) also noted that the position of the fins at the far end of the mantle allows them to be highly efficient rudders, since they can produce a much higher torque than would be possible if they were closer to the centre of dynamic pressure. This position far from the centre of gravity also makes them good stabilizers.

Although their fins are both good stabilizers and capable of undulatory propulsion, cephalopods have retained the jet, suggesting that fins and jet together are more

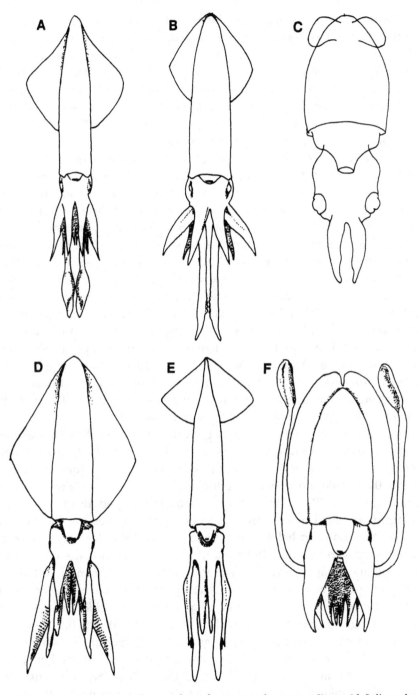

Figure 2. Cephalopod fin forms. Changes through ontogeny for a generalist squid, *Loligo vulgaris*: (A) adult (266 mm ML); (B) juvenile (81 mm ML); (C) hatchling (2·5 mm ML). Other specialized fin types: (D) *Thysanoteuthis rhombus*; (E) *Illex illecebrosus*; (F) *Sepia officinalis*. (Redrawn by J.A.A. Perez from Naef, 1921/23, except for (C) from Packard, 1969).

effective than fins alone, or that cephalopod undulatory fins are less efficient than those of fishes. Direct evidence for this is limited, but Webber (1985) noted that some individual *Illex illecebrosus* use fins more than others at low velocities and that those that seldom used fins generated higher mantle pressures than those which used them often at the same swimming speed. Both Webber (1985) and Packard (1969) also noted that squid use a strong fin beat during refilling, perhaps to lessen the detrimental effects of this stage of the cycle, to decrease the acceleration reaction by smoothing the overall velocity curve and to increase Froude efficiency.

Like the mantle (Otis & Gilly, 1990), squid fins are used in many different ways: at low speeds, undulatory fin waves produce thrust; at high speeds, they are rolled tightly against the mantle to decrease the drag; and at moderate speeds, the very action of rolling the fins around the mantle produces a small jet thrust (Clarke, 1988). Fin movement is usually synchronous on both sides of the mantle, but fins are capable of independent movement, as in pivoting cuttlefish, which produce waves in opposite directions on the two sides of the body. The velocity of the fin waves also varies with the velocity of the squid, from 4-5 strokes per minute, during immobile hovering (Zuev, 1966), to up to 90 beats per minute, during spawning when the jet cannot be used (O'Dor & Balch, 1985). Williamson (1965) noted that during escape movements, *Illex illecebrosus* synchronized the downbeat of the fins with the mantle contraction to maximize the total thrust production. When hovering, the fins produce synchronous waves from the mantle tip towards the head; this produces a propulsive force directed upwards and finwards along the longitudinal axis of the body, and therefore counter-acts the horizontal component of the jet thrust from the funnel, as well as providing lift to counteract the negative buoyancy of the squid (Zuev, 1966).

Lift generation is very important in negatively buoyant cephalopods as they will always sink unless something counteracts gravity. Even the body shape of cephalopods contributes to lift generation, since it has some vertical curvature and the animal swims with the mantle end raised to produce an angle of attack of 5° to 30°, depending on swimming speed (Zuev, 1966; O'Dor, 1988b). Most squid also have a horizontal thickening of the third arms, which function as fins, increasing the total surface extension of the body and helping to create a greater lifting force during swimming than would be possible without them (Zuev, 1966). The importance of undulatory fin thrust in counteracting negative buoyancy was clear in Williamson's (1965) ampu-tated, sinking squid. The lift from the fins is not enough to maintain an ommastrephid's position in the water column, however, as an adult female *Illex* using just her fins during spawning, continued to sink (O'Dor & Balch, 1985). The lift from the fins is essential for immobile hovering in squid and is very important for slow swimming (Zuev, 1966). At high speeds, when fins are wrapped around the body, all lift is generated from the body shape and the jet (O'Dor, 1988b). *Sepia* uses its fins much more than most cephalopods, probably due to its fin shape and shallow coastal life style, but the fins are not needed for lift generation since it possesses an internal shell to maintain neutral buoyancy.

Daniel's (1988) analysis of skate swimming using flapping fins found that hydrody-namic constraints are the major determinants of 'wing' shape, and that unsteady movement by low aspect-ratio fins may have a lower cost of transport than higher aspect-ratio wings moving the same way. The degree of unsteadiness and the number of waves along the wing are the important determinants of optimal wing shape for propulsion. Therefore one expects that for slow wing oscillations the higher aspect-ratio wings will produce fewer waves along them, while lower aspect-ratio wings will produce more waves at higher wing oscillations. Daniel (1988) found that manta and eagle rays fly with higher aspect-ratio fins and fewer propulsive waves than skates with lower aspect-ratio wings. Blake (1983b) found that knife fishes and electric fishes, with their very low aspect-ratio fins, passed multiple propulsive waves along their length, probably using balistiform locomotion (in which waves are propagated simultaneously along paired, flexible fins on a relatively wide, rigid body; Lighthill & Blake, 1990). In cephalopods, the low aspect-ratio fins of the sepioids produce multiple propulsive waves, while the more triangular, higher aspect-ratio fins of oceanic squid (e.g. ommastrephids) tend to produce fewer propulsive waves. When swimming normally, cuttlefishes propagate waves simultaneously along paired, highly flexible fins while maintaining a fairly rigid body, consistent with true balistiform swimming, which is probably the lowest cost method of swimming (Lighthill & Blake, 1990). Therefore, the natural range of fin shapes and distortions in both fish and cephalopods seems to mirror the full range of theoretical systems.

Some squid are really just aquatic flyers, and in flight aerodynamics Weis-Fogh (1973) and Lighthill (1973) described a system whereby insects could generate extra lift by 'clapping and flinging' their wings at the top of the wing-beat cycle, through the generation of vortices around the wing tips as they separated during the 'fling' movement. Lighthill (1973) even went further to suggest that if an insect could continue the wing movements through 180° and produce the clap and fling motion at both ends of the wing beat, then it would get an even greater advantage from lift generation at two points within the cycle. Some squid appear to clap their fins together at the top and possibly the bottom of the flapping stroke at high speeds making it possible that these animals are taking advantage of the Weis-Fogh mechanism to help counteract negative buoyancy. To date, no one has examined this potentially useful mechanism in cephalopods, or any other aquatic flyer.

A number of oceanic squid, especially the ommastrephids, will actually fly in air when chased (Cole & Gilbert, 1970; Murata, 1988). They have, therefore, been likened to flying fish that escape the water's higher drag to decrease their cost of transport. Lane (1960) reports estimates of squid flying 6 m up in the air and covering 20 m horizontally. Flying squid, however, differ from flying fish in that a school of squid continues to move in formation out of the water (even falling back into the water together), whereas the flying fish act as individuals and do not maintain the formation of the school throughout flight (Lane, 1960). Interestingly, the fin area of these flying squid is very small in comparison to that of flying fish (Packard, 1972), and could not produce much lift. In fact, many flying squid roll their fins, so that both acceleration and lift probably come from the continued expulsion of water during flight (Cole & Gilbert, 1970).

SCALING OF FINS

Most squid hatch with small paddle-shaped fins and a saccular body, quite different from the long, streamlined body of the adults (Figure 2; Zuev, 1964; Clarke, 1966). Squid hatchlings are immediately capable of independent movement, including co-ordinated jet locomotion (Packard, 1969), but the full flexibility of routine and escape jets is lacking (Gilly et al., 1991). Hatchling squid swim at moderate costs and velocities averaging 10 mm s^{-1}, but their lateral movement is limited (O'Dor et al., 1986). Although reasonably proficient at jetting, they have very little control, probably as a result of the small size of the fins. Boletzky (1987) noted that during the crucial pouncing movement of prey capture, the tiny fins of many squid hatchlings would probably be unable to stabilize the animal. He hypothesized, however, that the globular form of the mantle and the relatively large funnel might help the squid, due to the alignment of the centres of thrust and drag in forward movement. Therefore stabilization could be a function of the general body shape until the fins reach a useful size. Since the fins are so small at this stage, they also probably have little effect on lift generation, even though the hatchlings are more negatively buoyant than the adults, as in *Loligo vulgaris, Symplectoteuthis oualaniensis, Ancistroteuthis lichtensteini* and *Illex coindetii* (Zuev, 1966). This may be compensated for by the more vertical orientation of hatchlings (O'Dor et al., 1986).

To date, studies of the mechanics of cephalopod hatchling swimming have been very limited (Packard, 1969; O'Dor et al., 1986), largely because it traverses the range of Reynolds numbers (1 to 1000) of mixed viscous and inertial forces which are poorly understood in terms of drag estimations (Blake, 1983a). This means that studies of scaling in locomotion have to take both sets of forces into account, using inertial theory to describe the motion in the initial part (*i.e.* 200 ms) of a swimming event and then resistive theory afterwards (Daniel et al., 1992), since thrust is dominated by resistive forces after the inertial start-up. Although the theoretical approach is quite difficult, studies of swimming in hatchling and juvenile cephalopods are an important challenge, particularly because the allometry of the fins differs from that of other structures.

In the absence of theory, O'Dor & Webber (1986) noted that available data suggested a difference in the scaling of cost of transport between jet and undulatory propulsion. For squid in the size range 0·03 to 3 kg, cost of transport scales approximately as mass$^{-0.2}$, while for undulatory fishes the factor is mass$^{-0.3}$. Projected to smaller sizes, these factors predict that jetting is more efficient than undulating below a certain size. Consequences for the competition between squid and fish were suggested, and the ontogeny of squids appears to reflect the effect as well, as indicated in Figure 2. This trend is also seen in the adult loliginids which mature at different sizes, as shown in Table 2 (fin-length:mantle-length ratio increases from 0·47 to 0·65 from the smallest *L. opalescens* to the largest *L. forbesi*). Maximum fin-wave velocity also increased with squid length from 0·11 to 0·26 m s^{-1}; these velocities are well below those for similarly sized fishes. Models of the production of waves by muscular-hydrostats suggest that there may be a harmonic component (Johnsen & Kier, 1993), but to date it remains

Table 2. *Loliginid fin dimensions, velocities and forces.*

Ontogenetic changes in fin/mantle length ratios

Mantle length (mm)	L. opalescens[a]	L. pealei[b]	L. vulgaris[c]	L. forbesi[d]	
2.0		0.16			
2.3			0.21		
2.5			0.28		Hatchling
3.5	0.17				
4.9				0.22	
6.3			0.27		
8.0	0.15				
17	0.23				
24	0.36				
31		0.39			
34	0.39				
37			0.51		
58	0.45				
88			0.68		
95	0.47				
150			0.69		
196			0.71		
216				0.72	
220			0.70		Adult
332		0.68			
600				0.65	

Fin wave velocities and forces for adults in swim-tunnels

	L. opalescens	L. pealei		L. forbesi
Maximum current for continuous fin waves (cm s^{-1})	10[e]	14[f]		26
Average fin wave speed at current speed	9	14		17
Fin force (N)	0.004	0.029		

a, Unpublished growth series from Hanlon *et al.* (1979); b, Haefner (1964); c, Naef (1921/23), Packard (1966); d, Segawa *et al.* (1988) and unpublished data; e, O'Dor (1988b); f, Sim (1988).

unclear whether increasing wave velocities reflect simple mechanical factors, such as increasing fibre length in larger, thicker fins, or actual ontogenetic changes in shortening rates, as have recently been demonstrated in fishes (Altringham & Johnson, 1990b). However, the muscular hydrostat systems in squids lack the potential to multiply muscle shortening speeds through leverage using rigid skeletal elements, such as fin rays, which are available to fishes. This may be the ultimate constraint on cephalopod fin use.

FIN-JET INTERACTIONS

O'Dor (1988) estimated the fin contribution to swimming forces by difference for 0·03-kg *Loligo opalescens* at 0·004 N and suggested that the fins contributed thrust up to 0·2 m s^{-1}, but no detailed analysis of swim-tunnel fin kinematics for squid has been published. Figures 3 and 4 illustrate the kind of data available from good-quality lateral video-recordings of a 0·3-kg squid, including animal velocity, velocity of the fin

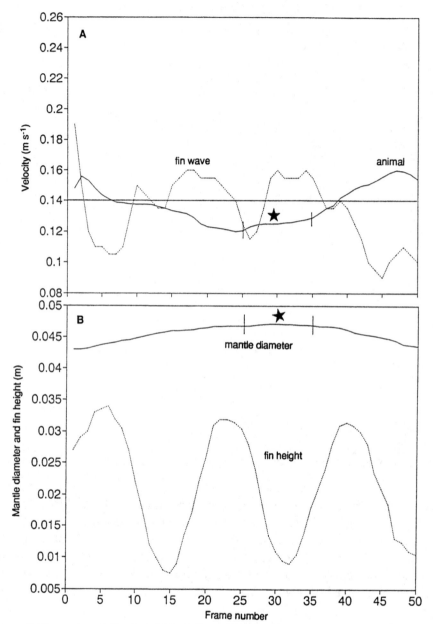

Figure 3. Kinematic variables for a 0·3-kg *Loligo pealei* swimming in a swim-tunnel with a current speed of 0·14 m s^{-1}. Decreasing mantle diameter (B, solid line) produces the jet which accounts for most of the acceleration in the cycle; refilling causes deceleration. Whole animal velocity (A, solid line) is calculated from movements of the mantle tip relative to the camera frame, allowing for constant current velocity. Mantle diameter is taken mid-way along the mantle at the widest point, and fin-wave height (B, dotted line) is measured at the same point. Wave velocity (A, dotted line) is calculated independent of current velocity from the movement of a point of maximum excursion as it passes along the fin. Fin thrust was calculated for the stretch of video during which mantle volume is constant, indicated by * (see text). Film speed=30 frames s^{-1}. The horizontal line in A indicates the water current velocity.

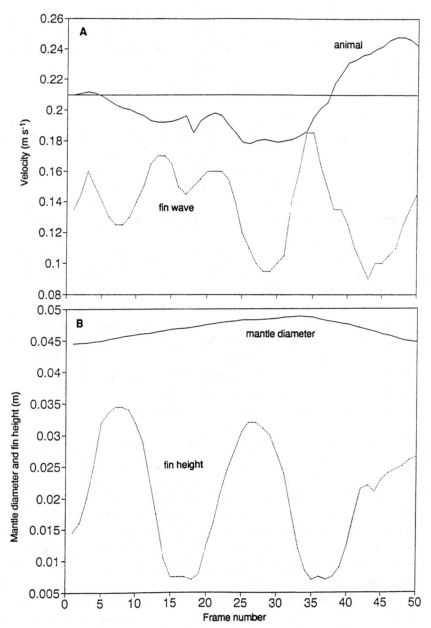

Figure 4. Kinematic variables as in Figure 3 at 0·21 m s⁻¹.

waves, wave amplitude and mantle diameter. Values for these variables are shown for 0·14 and 0·21 m s⁻¹, averaged over five jet cycles from frame-by-frame analysis of *L. pealei* (Sim, 1988). At a current velocity of 0·14 m s⁻¹ animal velocity fluctuates from 0·12 to 0·16 m s⁻¹ and wave velocity averages 0·14 m s⁻¹, ranging from 0·11 to 0·19 m s⁻¹. There are three full waves per jet cycle. Over the period indicated with stars, there was no net change in mantle volume, and calculations like those used by Blake (1983b) indicate that fins produce a thrust of 0·029 N, yielding a slight acceleration from 0·122 to 0·128 m s^{-1}. The average and range of wave velocities are the same at 0·21 m s^{-1}, but because these squid swim more steadily (velocity range is 0·18 to 0·25 m s^{-1} compared with 0·12 to 0·31 m s^{-1} for the small *L. opalescens* at 0·20 m s^{-1}), there appears to be no contribution to thrust by the fins at any point in the cycle. Fin waves continue to much higher velocities, but their frequency decreases, as can be seen in Figure 4B, indicating that they are too slow to produce thrust. The thrust estimate at 0·14 m s^{-1} appears reasonable in comparison to *L. opalescens* and to finning fish, but could be checked in detail. Pressure data are available for a force analysis (O'Dor, 1988a), and the kinematic data would allow more sophisticated analyses of undulatory fin thrust (Daniel, 1988; Lighthill & Blake, 1990). This should be an interesting case for comparison of jet thrusts with those predicted by new theoretical approaches to the thrust from undulatory waves.

Squids commonly make a manoeuvre rare in fishes, a complete reversal of direction without a turn, which provides another opportunity to test theory. Foyle & O'Dor (1988) describe attacks on small fishes in which a squid overtakes a school of fish by swimming fins-first, and then lunges back into the school head-first to make a capture. In the 15-m diameter pool (Webber & O'Dor, 1986), long-finned squids also hold position by repeatedly reversing along their horizontal axis, while short-finned squids more commonly bob up and down as they jet. The large, long-finned *L. forbesi* studied in the Azores (O'Dor *et al.*, 1993b) used this manoeuvre to avoid crashing into the walls of their 3-m tank. With simultaneous top-down video and telemetered jet pressures available, this provided an unusual opportunity to look at the interaction of fins and jets.

Figure 5A shows a pressure record for a 2-kg squid carrying a transducer that transmits a pulse every 2 s at 0 kPa, and decreases its pulse interval to 0·5 s at 40 kPa. The squid goes through a complete reversal cycle, starting out fins-first towards a wall, reversing to head-first towards the opposite wall and then reversing to fins-first to return. Figure 5C shows the corresponding animal and fin-wave velocities. Fins-first movement of the animal is taken as positive velocity, but the polarity of wave velocity is reversed, as is customary in dealing with fish in which motion is only possible when waves produce forward momentum by pushing water backwards. The situation is much more complex in squids where wave direction can differ from animal direction. Lighthill's (1971) calculation of Froude efficiency for undulatory swimming (wave velocity, *c*, plus animal velocity, *u*, divided by 2*c*) probably does not really apply in such circumstances, but it gives an indication of what is happening in Figure 5B.

In fins-first swimming at 0·2 m s^{-1} the fins function very efficiently and the waves reverse direction coincidentally with the finward jet; this rapidly slows the animal down until the fins reach full efficiency in head-first swimming. Then, half-way across the pool, the squid, perhaps seeing the wall approaching earlier in this direction, reverses its fin waves, but nothing happens! There is no change in animal velocity. Is Froude efficiency really zero? At the last moment, a headward jet, of one-quarter the pressure of the earlier finward jet, reverses the direction and rapidly accelerates the

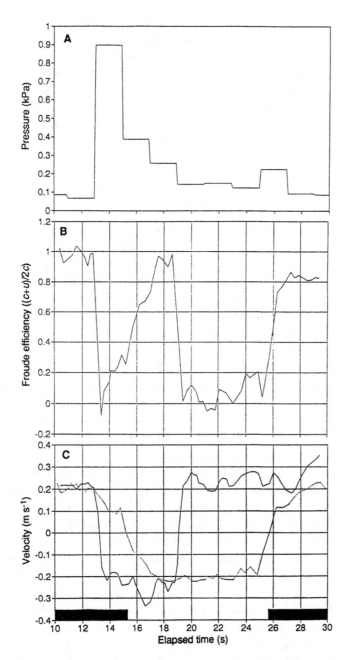

Figure 5. Simultaneous records of jet pressure, animal velocity and fin-wave velocity during a typical cycle of fins-first and head-first reversals by a *Loligo forbesi* in a 3-m tank. (A) indicates average pressures over intervals of about 2 s from ultrasonic telemetry. In (C) animal velocity (*u*) is the solid line, wave velocity (*c*) the broken line, and the black bars indicate that the squid is swimming fins-first. 'Froude efficiency' in (B) is calculated as $(c+u)/2c$. (See text for details.)

squid to 0·2 m s⁻¹, where the fins again become briefly effective. We suspect that this apparently uneconomical behaviour is an adaptation of the to-and-fro hovering pattern of long-finned squids mentioned above. In less confining circumstances (such as the open ocean or a larger tank) the fins-first swim would be directed upward so that negative buoyancy would slow the squid to zero before the second jet and fin reversal occurred; it is easier to stop in the head-first direction by spreading the arms to increase drag. Although the behaviour recorded in Figure 5 seems to waste energy, it is preferable to crashing into a wall, and most squid learned to modify their behaviour in a few hours.

FUTURE DIRECTIONS

Cephalopods have developed complex locomotory systems in their fight to retain (and perhaps regain) their position in the oceans. Although jet propulsion has been fairly well studied, there is still much work required on the scaling of the jet and its use in young squid. Cephalopod fins, on the other hand, have barely been studied quantitatively as propulsive organs, even though they are used with the jet. With the increase in field tracking studies, it is important to understand the roles of the fins in locomotion in Table 1, but this is only the tip of the iceberg. The frequency of waves relative to jet frequency varies at different swimming speeds, producing what appear to be 'gaits' (O'Dor, 1988a). Some of these effects may be hydrodynamically coupled, but independent neural control must also be a factor. Short and even intermediate-length fins can be flapped as well as waved. At steady low speeds, waves continue throughout the jet cycle, while at intermediate speeds they occur only during refilling, and at high speeds the fins are only occasionally flapped to control angle and direction. When struggling to escape from a jig, *L. forbesi* synchronizes the down-stroke of the flap with the jet to maximize thrust. Even full-length fins appear to have different operating modes. *Sepia* seem to meet the criteria for being balistiform swimmers, perhaps accounting for their surprising quickness, while diamond-finned species like *Thysanoteuthis rhombus* are probably limited to the equivalent of less efficient rajiform swimming. Although the anatomical sources of the controls that produce and optimize all these different styles of locomotion are known (Boycott, 1961), we are only now beginning to appreciate their full complexity (Gosline *et al.*, 1983; Gilly *et al.*, 1991).

Several laboratory analyses mentioned suggest that squid fins are often a handicap, to be rolled up while exploiting the jet, but all this really proves is that squid did not evolve to live in laboratories. Squids are among the most powerful animals in the sea when applying their jets to escape or pursuit, and it is clear that fins are a detriment, not an aid to such activities. Squids with large fins waste power in rapid manoeuvres dragging their fins around, but for cheap transportation fins out-perform jets. Based on swim-tunnel studies, which are artificial paradigms, the minimum cost of transport for a 0·6-kg *Sepia* is 3·2 J kg⁻¹ m⁻¹ at 0·3-0·35 m s⁻¹ (O'Dor & Webber, 1991). Although twice the cost for a similar-sized salmon, it is only 67% of *Loligo* and 40% of *Illex* costs at the same velocity. Figure 1 suggests that at low speeds the jet may be the handicap.

On the other hand, the results in Figure 5C suggest that even low-level jetting associated with respiration may help to keep animal velocity and fin-wave velocity closely matched, thus maximizing Froude efficiency. At low speeds, there appears to be a potential for synergy between respiratory pumping and wave propulsion which could increase efficiencies for both jet and fins, but may be difficult to analyse given the difficulty of interpreting low-speed respirometry in swim-tunnels (O'Dor, 1988a).

It is evidence of their remarkable neural flexibility that these animals manage the kind of experimental paradigms we set for them at all; we should not be surprised if they are not very efficient at them. To understand fully what adaptations suit animals to their life-styles requires knowing what those life-styles are. Various squid designs must represent a wide array of compromises and synergies. While it might be possible to work out what various species are best at in the laboratory, it is probably easier, given the available techniques for tracking and monitoring jet pressure (Webber & O'Dor, 1986; O'Dor *et al.*, 1993a,b), to find out what they actually do in nature and then compare efficiencies at these tasks experimentally. Ultimately, all our conclusions about locomotor adaptations need confirmation in nature; why not do it first? This remains the weakest link in most studies of locomotion; the ease with which jet power can be monitored should make it a useful model for evaluating the natural application of other types of locomotor studies.

Undulating squid fins provide more efficient transport than the jet, but squid fin undulations do not appear to be directly comparable to fish fin or whole-body undulations. Squids use finning as an economical alternative to jetting, but only at low speeds. Many squids may depend on their fins for a greater proportion of their total locomotion than the jet, but the jet is essential for high-speed manoeuvres and may enhance the efficiency of fins. It is unclear whether the production of undulations by muscular hydrostats is more or less efficient than by musculo-skeletal systems, but it appears to be slower. There may be nothing that a squid without a jet could do better than a fish, which would account for the fact that no cephalopod has abandoned jetting completely (Wells & O'Dor, 1991). Squid may be evolving toward faster muscles, but, in the meantime, there are at least a few fish experimenting with jets (Fish, 1987).

Basic supported was from grants to R.K.O. from the Natural Sciences and Engineering Research Council of Canada and an NSERC Postgraduate Scholarship to J.A.H. The work on *Loligo forbesi* was funded by grants to F.G. Carey, WHOI, from NSF and the National Geographic Society and that on *Sepia officinalis* by grants to Phil Lee, MBL, from NATO.

Chapter 4

The biology of fish swimming

PAUL W. WEBB

School of Natural Resources & Environment and Department of
Biology, University of Michigan, Ann Arbor, MI 48109-1115, USA

This chapter explores the ways fish swim from zero speeds in station-holding and hovering, through cruising and sprint, to fast starts. The range of power required to swim over such a range is formidable. Effective swimming is achieved by performance range fractionation using gaits. Gaits are defined by the use of various combinations of propulsor type (median and paired fin or body / caudal fin propulsors), propulsor kinematics (station-holding, hovering, steady swimming and fast starts, muscle (red, pink and white), and locomotor behaviour (continuous and burst-and-coast swimming). The number of gaits expressed within lineages, and presumably locomotor performance range, has generally increased over evolutionary time. At any given evolutionary level, radiations have been common, often enhancing certain gaits but reducing performance in others (gait suppression). The range of gaits expressed during ontogeny also increases. Environmental influences tend to reduce metabolic scope so that successive gaits tend to be recruited at lower speeds (gait compression). Directions for future research include use of computational models better to model pressure distributions and forces during swimming. Problems especially of manoeuvrability, agility and stability have been neglected by traditional interest in maximum speed and acceleration.

INTRODUCTION

Fish are famed for their swimming prowess. But many species also walk over the aquatic and terrestrial substratum (Grobecker & Pietsch, 1979; Gans et al., 1994) and others glide through the air (Fish, 1990). Fish swimming immediately suggests flashing fins, but some move by jet propulsion (Fish, 1987). Thus fish employ all axial and appendicular propulsor organs, their attendant muscles, and several behaviours in a wide variety of locomotor modes (Webb & Blake, 1985; Webb, 1993, 1994a).

Maddock, L., Bone, Q. & Rayner, J.M.V. (ed.). Mechanics and Physiology of Animal Swimming.
© 1994. Cambridge University Press.

Here, I focus on swimming by using the concepts of gaits to explore the *intraspecific* question of how fish swim effectively over a formidably large range of speeds and accelerations. This differs from traditional classifications based on *interspecific* differences in propulsor morphology, propulsor kinematics, and momentum transfer mechanisms (Breder, 1926; Lindsey, 1978; Blake, 1983a; Braun & Reif, 1985; Webb & Blake, 1985; Daniel & Webb, 1987; Webb, 1988). The most popular traditional classification builds on the primarily morphological approach of Breder (1926). He used exemplary groups for particular propulsors, adding the suffix 'form' to define swimming patterns in a way that readily calls to mind a visual image of a swimming style (Figure 1). For this reason, Breder's nomenclature remains useful. However, no satisfactory unified classification of fish swimming modes has emerged.

The concept of gaits, derived from Alexander (1989a), provides a framework inclusive of traditional mechanical, physiological and performance aspects of swimming, including environmental effects. Variations among species are incorporated for ontogenetic, ecological and evolutionary perspectives. The approach leads to recognition of special problems associated with stability, manoeuvrability and agility which have previously been neglected. Many areas are revealed for which we lack data. Hence many suggestions are necessarily speculative and many will be refined, rejected or succeeded in the light of better information.

GAITS AND SWIMMING PERFORMANCE RANGE FRACTIONATION

Basic concepts

Definition

The concept of gaits, such as walk, trot and gallop, derives from studies of pedestrian locomotion in which foot-fall patterns change discontinuously as speed increases (Hildebrand, 1985; Bennett, 1992). Alexander (1989a) proposed a definition of a gait as "a pattern of locomotion characteristic of a limited range of speeds described by quantities of which one or more change discontinuously at transitions to other gaits". This definition is applicable to pedestrians, swimmers and flyers.

Energy minimization

Pedestrian gaits are viewed as mechanisms dividing an animal's performance range into regions that minimize the work required of the muscles at each speed (Hoyt & Taylor, 1981; Alexander, 1984; Bennett, 1992; Casey, 1992). By analogy with concepts from sensory physiology, this can be viewed as performance range fractionation using gaits (Webb, 1993, 1994a).

Startle responses

Increasing speed is not the only factor associated with transitions in locomotor patterns, and efficiency criteria are not sufficient to understand gait transitions (Bennett, 1992; Webb, 1994a). The final defence of animals in critical situations, such as attack by a predator, is the startle response. This is a unique locomotor pattern characterized by high acceleration rates in which survival takes precedence over efficiency (Eaton & Hackett, 1984; Feder & Lauder, 1986).

The startle response, or a fast start in fish, is the large amplitude non-repeated movement comprising the first tail-beat following a startle stimulus. The aperiodic movement lasts from about 60 to 150 ms (depending on fish size) and is driven by unilateral contractions of the myotomal muscle on each side of the body. Rates of acceleration typically peak during this period (Weihs, 1973b; Webb, 1993, 1994a).

Gaits of swimmers

Alexander (1989a) emphasized that the principles defining gaits are globally applicable to animal locomotion (Rayner et al., 1986; Alexander, 1989a; Webb, 1993, 1994a). Applying his definition of a gait to fish, however, encompasses a wider range of components than are typically used to differentiate pedestrian gaits. The wider range of gaits reflects problems of moving through fluids.

Propulsors

First, swimmers differ fundamentally from terrestrial animals because most of the body weight is buoyed up by water upthrust (Alexander, 1990). Since weight support is not a dominant problem, numerous propulsive systems, which can function independently or together, have evolved among fish (Daniel & Webb, 1987; Webb, 1994a). As a result, fish not only can change gaits by moving a given propulsor with a distinct kinematic pattern, but also recruit separate, and hence kinematically distinct, propulsors (Alexander, 1989a; Webb, 1993, 1994a).

Slow speed swimming tends to be powered by median and paired fins (defined as MPF swimming), supplemented and then replaced by body and caudal fin undulation (defined as BCF swimming) at higher speeds and for high rates of acceleration (Figure 1).

Hovering and station-holding

Pedestrian gaits apply to animals moving at finite speeds relative to the ground or the air. Velocity, u, (or for practical purposes, speed), is one parameter defining the state of a body, and hence u may equal zero. At zero water-speed fish hover, using distinct movement patterns that therefore define a hovering gait (Alexander, 1989a).

Zero ground speed is difficult to achieve in water currents, a disadvantage of the buoyant effects of water (Daniel & Webb, 1987; Full & Koehl, 1993). The advantages of high rates of food delivery, gamete dispersion etc. are offset by limited performance capacity and high swimming costs, and exploitation of current-swept habitats typically requires minimizing or avoiding swimming. Fish hide from flows using habitat structure as refuges: substratum and protruding ripples, rocks, macrophytes, coral and submerged rootwads and branches. Because of differences in the force balance for fish at the substratum vs swimming in the water column (Daniel & Webb, 1987), flow avoidance is

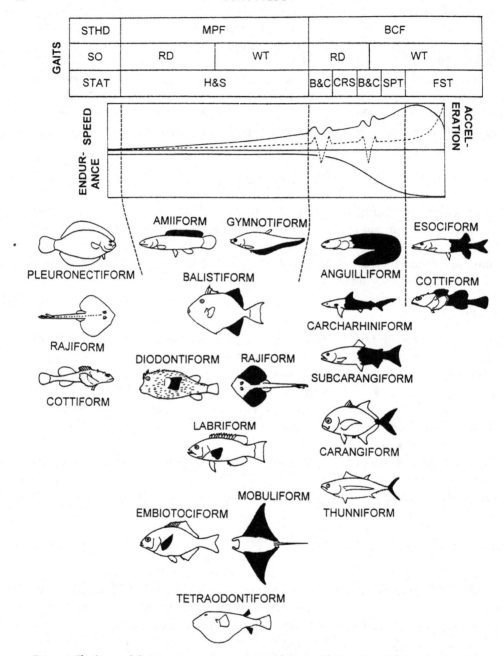

Figure 1. The factors defining gaits are summarized in the box at the top. Here fish may hold station (STHD), use median and paired fin propulsors (MPF), or body and caudal fin propulsors (BCF). Various muscles may be used, red (slow oxidative) or white (fast glycolytic); fish have little pink (fast oxidative glycolytic) muscle, which has been omitted for convenience. Various propulsor-muscle combinations may also be used in kinematically distinct behaviours. Station-holding fish are stationary relative to the substratum (STAT). Fish using MPF gaits typically hover and swim slowly (H&S). Some fish using red muscle in BCF swimming may first show burst-and-coast kinematics (B&C) before cruising (CRS). Burst-and-coast swimming occurs again at prolonged speeds where white muscle is recruited. Higher speeds are sprints (SPT), while maximum acceleration is seen in the fast-start gait (FST). In the upper part of the central panel trends in speed (solid line) and acceleration

facilitated by morphologies and behaviours that are recruited discontinuously at zero ground speed and which are distinct from those facilitating translocation. Therefore, station-holding is an important energy-minimizing gait for fish.

Muscle

Numerous studies have shown that different muscle systems are recruited, often discretely and discontinuously, as swimming speed increases (Bone, 1966; Johnston *et al.*, 1977; Goldspink, 1981; Rome *et al.*, 1990; Johnston, 1991; Rome, chapter 6). At lower body and caudal fin swimming speeds, slow oxidative (red) fibres provide the propulsive energy. These non-fatiguing fibres do not have a high power output and propel fish at low sustainable cruising speeds, defined for practical purposes as speeds sustained for >200 min (Hoar & Randall, 1978). At higher swimming speeds, fast oxidative glycolytic (pink) and/or fast glycolytic (white) fibres are recruited. The white fibres have a high power output but limited energy capacity (Bennett, 1985, 1991; Rome *et al.*, 1988). As a result, fish eventually fatigue at these speeds, defined therefore as prolonged speeds maintained from 15 s to 200 min (Hoar & Randall, 1978). The highest speeds are sprints, lasting less than 15 s (Hoar & Randall, 1978). Power for sprints and fast starts comes primarily from myotomal white muscle.

The change from red-fibre based body undulations to white muscle swimming is discrete in elasmobranchs, sarcopterygians and less derived teleosts (Halecostomes, Ginglymodi, basal elopomorphs and clupeomorphs) with focally innervated white fibres. In more derived teleosts with multiterminal innervation of white fibres, graded responses are possible and red and white fibres may overlap in their use at intermediate swimming speeds (Bone, 1978; Fetcho, 1987). Overall, muscle types are recruited for portions of the total performance range in a sequence which matches power and efficiency to the requirements for overcoming resistance in each portion. Hence muscle use is a quantity that varies with speed, meeting the criteria for distinguishing gaits (Alexander, 1989a).

Median and paired fin muscle has not been as well studied as myotomal muscle, but red, pink and white fibres have been identified. Sequential red, pink and white recruitment is possible, but in contrast to myotomal muscle, red fibres are most abundant (Davison & MacDonald, 1985; Hochachka & Somero, 1984; Archer & Johnston, 1989). White fibres may be more important to adduct the fins quickly, to reduce drag in sprints and fast starts (Harrison *et al.*, 1987; Johnston, 1991).

(dotted line) through various gaits are illustrated. In the lower central panel endurance patterns are illustrated. The lower part of the figure shows external morphologies and nomenclature in the popular interspecific classification of propulsors (shown shaded), based on Breder (1926), Lindsey (1978) and Webb (1982, 1984). Modes are aligned in columns beneath the appropriate STHD, MPF, BCF steady swimming and BCF fast-start gaits. The named swimming modes are typically specializations to improve motor performance in certain gaits. The traditional ostraciiform mode is omitted. Blake (1977) clearly shows that the caudal fin movements for which the mode is defined are really a component of MPF swimming. MPF and BCF gaits are ranked from top to bottom in order of increasing motor performance, associated with shifts from undulatory resistance-based (top) to oscillatory lift-based (bottom) propulsion.

Burst-and-coast behaviour

Alexander (1989a) pointed out that some kinematically distinct locomotor patterns span more than one stride, for example, bounding flight in birds and burst-and-coast swimming in fish. Burst-and-coast swimming alternates a brief period of swimming with a coast of constant depth or a downward glide (Weihs, 1973a, 1974).

Burst-and-coast behaviour is believed to increase endurance, being recruited at speeds where fatigue may occur. Low endurance is a potential problem at very low swimming speeds for fish that lack substantial median and paired fin swimming capabilities. These fish must use body and caudal fin gaits over almost the entire performance range. At low swimming speeds, the myotomal muscles would have to work at low strain rates at which efficiency is very small (Rome *et al.*, 1990). Overall energetic efficiency is promoted by allowing muscle to work at higher strain rates during the intermittent swimming phase. Burst-and-coast behaviour is also used at prolonged speeds where it can reduce energy expenditure by over 50% compared with continuous swimming at the equivalent mean speed (Weihs, 1974; Videler & Weihs, 1982). The energy saving occurs because the drag of the stretched-straight body in the coasting phase may be 3- to 5-fold lower than that in the undulatory body and caudal fin swimming phase (Alexander, 1967; Lighthill, 1975; Webb, 1975a).

The general gait recruitment sequence

I propose that gaits in fish are defined by the following quantities (Figure 1); (a) propulsor use (MPF and BCF swimming); (b) kinematics (station-holding, hovering, steady swimming and fast starts); (c) muscle fibre use (red, pink and white); (d) locomotor behaviour (steady and burst-and-coast). Various combinations (gaits) of these quantities are recruited in an orderly sequence as speed increases from zero to maximum bursts.

The lowest levels of performance are zero ground-speed in currents, and zero water-speed hovering using MPF gaits (Figure 1). Both are sustainable activities, presumably supported by red muscle.

Translocation at the lowest swimming speeds also uses MPF propulsors and probably red muscle. Paired pectoral and pelvic fins and median anal, dorsal and caudal fins are arranged about the centre of mass, where they most commonly work together to stabilize slow swimming. Red muscle is probably the major source of motive power, but an MPF-white muscle gait cannot be ruled out.

As speed increases, MPF propulsion is first supplemented and then succeeded by BCF propulsion. At the lowest BCF swimming speeds, the propulsor may be driven by myotomal red muscle in a burst-and-coast gait (Rome *et al.*, 1990). Fish cruise at higher speeds using BCF-red muscle gaits before again using the burst-and-coast gait at prolonged speeds (Videler & Weihs, 1982; Alexander, 1989a).

With further increases in speed, pink and white muscles supplement and then succeed red muscle. Maximum sprint speeds are powered by myotomal muscle, which, because of the high power requirements for fast swimming, may represent as much as 65% of the body mass (Bainbridge, 1961).

Successive gaits probably provide increased acceleration performance in parallel with higher speeds. In a turn, the force required for centripetal acceleration increases with u^2/r, where r is the turning radius. Since r is independent of u (Howland, 1974; Webb, 1983), forces and power required for turning follow those for speed.

The forces for linear acceleration and deceleration are proportional to du/dt. Low rates are readily incorporated into routine swimming in most gaits through local changes in propulsor beat period, amplitude or orientation. Maximum accelerations are achieved in the fast-start gait.

Why do fish have so many gaits?

Fish can express a large number of gaits. This may be essential if they are to swim over a large range of speeds and accelerations. The power for locomotion in terrestrial pedestrians increases linearly with speed, and in flyers, power is related to speed by a U-shaped function which can be shallow enough to approach speed independence (Norberg, 1990; Casey, 1992). Mechanical power of swimming in fish increases with the 2·5 to 2·8 power of u. Thus the power requirements for swimming increase more rapidly with performance level than they do for terrestrial animals. Consequently a given gait in fish can efficiently power only a small range of speeds compared with a pedestrian or flyer. Therefore, a wide overall performance range for swimmers can only be achieved by a large number of gaits.

VARIATION IN GAIT EXPRESSION AMONG FISH

Evolutionary trends

The number of gaits used by a species, and hence its performance range, depends on the presence and developmental level of the propulsors, muscles and behaviours. The number of gaits has increased through evolutionary time (Figure 2). At the same time, there have been repeated radiations of fish at successive evolutionary levels into various habitats in which enhanced performance in one or more gaits has been favoured (Webb, 1982). Such specialization for a particular gait is usually associated with reduced performance in other gaits, and/or a reduction in the total number of gaits, which I define as gait suppression.

Traces of fish are found in the Ordovician, and fossils showing a body form similar to modern fishes become common in Silurian strata (Moy-Thomas & Miles, 1971). The earliest fish were armoured and had a tapered elongate tail. They are postulated to have swum very much like their chordate ancestors, and were probably restricted to BCF gaits. Armour would have promoted station-holding for stream and estuarine species, as well as for those exposed to tidal currents (see Moy-Thomas & Miles, 1971). On the other hand, armour adds substantial resistance to acceleration (Webb & Smith, 1980; Webb et al., 1992) so that the fast-start gait was probably missing, or at most poorly developed. The expansion into pelagic habitats by pteraspids and cephalaspids was

Evolutionary Trends in Fish Gaits

Palaeozoic fishes

Early Devonian and Silurian heterostracians, pteraspids and cephalaspids.

STHD	BCF			
RD	RD		WT	
STAT	B&C	CRS	B&C	SPT

Later Silurian pteraspids and cephalaspids with reduction or loss of armour.

STHD	BCF				
RD	RD		WT		
STAT	B&C	CRS	B&C	SPT	FST

Carboniferous fishes

STHD	MPF		BCF				
RD	RD	WT	RD		WT		
STAT	H&S		B&C	CRS	B&C	SPT	FST

Generalized teleosts

Malacopterygians - soft-rayed fishes.

STHD	MPF		BCF				
RD	RD	WT	RD		WT		
STAT	H&S		B&C	CRS	B&C	SPT	FST

Acanthopterygians - spiny-rayed fishes.

STHD	MPF		BCF				
RD	RD	WT	RD		WT		
STAT	H&S		CRS	B&C	SPT	FST	

Elasmobranchs

Selachians

STHD	BCF			
RD	RD		WT	
STAT	B&C	CRS	B&C	SPT

Batoidimorphs

STHD	MPF					
RD	RD	WT				
STAT	H&S					

Figure 2. A summary of major trends in gaits expressed during the evolution of fishes. STHD, station holding; MPF, median/paired fin propulsors; BCF, body/caudal fin propulsors; RD, red, slow oxidative muscle; WT, white, fast glycolytic muscle; STAT, stationary on substratum; H&S, hovering and slow swimming; B&C, burst-and-coast swimming; CRS, cruising; SPT, sprinting; FST, fast start. Probable differences in relative performance are illustrated by box borders. Lowest performance is suggested by single, narrow borders, with increasing performance indicated by narrow, double and then thick, double borders. Highest performance is indicated by a single, thick border.

associated with a reduction in armour together with increased thrust from a larger tail. The fast-start gait was probably added at this time (Figure 2; Webb & Smith, 1980). Muscle fibre composition is unknown, but was probably made up primarily of red and white types, similar to that in modern cephalochordates and agnathans (Bone, 1978).

Compared with modern fish, performance of Palaeozoic fish was probably not high in any gait. The notochord was unrestricted and vertebral elements strengthening the axial skeleton were poorly developed (Greenwood *et al.*, 1966; Schaeffer & Rosen, 1961; Schaeffer, 1967; Patterson, 1968a,b; Moss, 1977; Romer, 1977; Symmons, 1979; Webb, 1982). Elongate median fins and appendages were common (Moy-Thomas & Miles, 1971), but these lacked mobility and hence probably functioned mainly to trim BCF swimming and coasting.

Effective MPF gaits probably arose by the Carboniferous (Figure 2), when all the major groups of fish were represented. MPF gaits probably arose via movable stabilizers for trimming, which subsequently acquired the ability for independent movement (Weihs, 1989; Alexander, 1990). Internal pectoral and pelvic girdles appeared supporting fins on smaller bases (Lauder & Liem, 1983). In addition, spines usually supported the leading edge of fins. Median fin spines were firmly embedded in muscle. Except for the Crossopterygii, fins had less bulky supporting structures (Moy-Thomas & Miles, 1971).

Subsequent evolutionary patterns differ between the actinopterygians and elasmobranchs. Actinopterygian evolution is characterized by an increasing range of gaits while elasmobranchs use a limited range of gaits with substantial gait suppression.

Actinopterygian performance probably improved in all gaits, with continuous increases in strength of the axial and appendicular skeleton (Moy-Thomas & Miles, 1971; Moss, 1977; Compagno, 1977, 1988) and increasingly flexible rayed fins (Lauder & Liem, 1983). A generalized modern soft-rayed fish (e.g. elopomorph, clupeomorph) has well-developed BCF gaits, but, compared with acanthopterygians, has less developed MPF gaits. Malacopterygians use their ventrally located pectoral and pelvic fins (Rosen, 1982) to swim at very slow speeds, or as hydrofoils to avoid swimming (Arnold et al., 1991; Webb, 1994a). In addition, slow swimming may involve the BCF burst-and-coast gait (Rome et al., 1990).

In more derived paracanthopterygian and acanthopterygian (spiny-rayed) fish, pectoral fins are lateral, and pelvic fins (when present) are anterior and ventro-lateral (Rosen, 1982). The general acanthopterygian arrangement of the pectoral, pelvic, dorsal and anal fins distributes substantial MPF propulsors around the centre of mass, presumably improving stabilized hovering, slow swimming and manoeuvring. This expectation is supported by the prevalence of these fish in highly structured habitats such as coral reefs and macrophyte beds (Bone & Marshall, 1982; Ehlinger & Wilson, 1988). Improved hovering and slow swimming obviates the need for the slow speed BCF burst-and-coast gait. Coupled with plesiomorphic BCF gaits, expanded MPF gaits maximize performance range options, and undoubtedly this has contributed to the successful radiation of the acanthopterygian fish.

Elasmobranchs as a group have tended to emphasize gaits for cruising (Figure 2). BCF gaits are used by selachians, and may include median fin interactions to enhance thrust and improve efficiency (Lighthill, 1975). MPF pectoral fin propulsion is found in batoidimorphs. Many elasmobranchs are negatively buoyant and rest on the substratum rather than swimming slowly, reducing their need for MPF gaits. The fast-start gait does not appear to be well developed, perhaps because the large size of elasmobranchs from birth provides predator protection (Webb & de Buffrénil, 1990).

Specialization and gait suppression

Various innovations in muscle, skeleton and propulsor morphology are seen throughout the evolution of fish. These undoubtedly increased thrust forces and hence speed and acceleration (Webb, 1982; Lauder, 1988). Successive morphological grades have radiated into a wide range of habitats, often with specialization enhancing performance in one or more gaits, but eliminating or reducing performance in others (gait suppression).

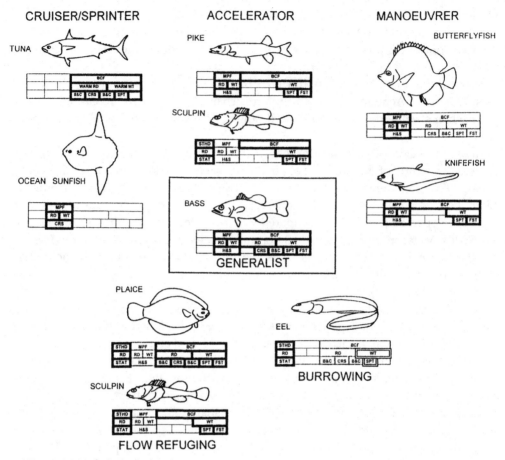

Figure 3. Morphologies and gait ranges of locomotor specialists. Differences in relative performance are illustrated by box borders, from narrow, double border for lowest performance through thick, double to single, double for highest performance. Muscle endothermy in tuna is indicated as warm red and white muscle. Other symbols are the same as in Figure 2.

Three types of specialization are associated with fish which live primarily in the water column (Figure 3): cruiser/sprinters, accelerators and manoeuvrers (Webb, 1982, 1984).

Marine pelagic fish tend to have enhanced cruising and sprinting gaits, travelling over wide areas in search of food and breeding grounds. Cruiser/sprinters most commonly propel themselves with thrust maximizing (Weihs, 1989), lift-based caudal fin propulsors (carangiform and thunniform mode), and occasionally with median fin propulsors (moliform mode). The optimal shape for such a propulsor is a narrow fin with a large span (high aspect ratio). The fin is typically stiff to resist span-wise buckling (Alexander, 1983; Magnuson, 1978; Hoerner, 1975). Cruiser/sprinters are often negatively buoyant and swim continuously, when hydrodynamic lift, commonly from stiff, high aspect-ratio, paired fins, is the most economical way to control position in the water column (Alexander, 1990). Stiff fins do not function efficiently in resistance-based oscillatory and undulatory modes characteristic of slow-speed swimmers, so that a general trend in cruiser/sprinters is MPF-gait suppression.

Slow-swimming ability is further reduced because cruiser/sprinters swim using long body waves (Wardle & Videler, 1980; Rome et al., 1992a,b; Rome, chapter 6). This reduces efficiency at low speeds (Rome et al., 1992a,b; Webb, 1994a). Negatively buoyant cruiser/sprinters are further constrained in their ability to swim slowly because there is a minimum speed at which lift can support their weight in water.

The fast-start gait is also suppressed in cruiser/sprinters (Webb & de Buffrénil, 1990) because lift is a surface force mis-matched to acceleration resistance, a volume force (Weihs, 1972, 1973b, 1974; Weihs & Webb, 1983). Endothermy, in which muscle temperature is regulated by metabolic heat produced in red muscles, may provide some compensation (Webb & de Buffrénil, 1990). Endothermy increases muscle power (Bennett, 1985, 1991) providing for higher speeds and hence higher lift forces for use in manoeuvres.

Finally, the pelagic life-style suppresses the station-holding gait.

Accelerators that enhance the fast-start gait tend to rest on the bottom or hover in the water column ready to ambush passing prey. These fish are characterized by a large caudal area due to a combination of a low aspect-ratio caudal fin, a deep caudal peduncle, and dorsal and anal fins located or extended posteriorly (Figure 3). The fin configuration ensures large masses of water are entrained along the body length to maximize thrust during fast starts (Weihs, 1972, 1973b). Two morphological variants may be recognized in pike-like (esociform) fish with a large caudal area and an elongated, circular anterior body, and cottid-like (cottiform) fish with a depressed body and a large fin depth along most of the body length. Enlarged pectoral fins may also contribute to acceleration of the latter (Montgomery & MacDonald, 1984). The differences are probably related to feeding habits (Webb, 1978a). Esociform fish have the highest acceleration rates recorded to date (Harper & Blake, 1990; but see Webb, 1994a).

Specialists in fast-start gaits that hover in ambush sites use MPF hovering and slow-swimming gaits, but speeds attained are very low (Webb, 1994a). Hovering and slow swimming are further reduced for fish striking from benthic ambush sites. Thus MPF gaits are poorly developed in accelerators.

In some ambushers, suction feeding replaces body translocation and all gaits are suppressed (Grobecker & Pietsch, 1979). Greatest suppression occurs in fish living in food-poor marine mesopelagic and deep benthic habitats. Low-thrust tapering body forms and reduced muscle are common (Bone & Marshall, 1982; Hochachka & Somero, 1984; Webb, 1990).

Manoeuvrers are found in two principal habitats. Many wind their way through complexly structured habitats such as coral reefs and macrophyte beds, orientating the body in many different directions to pick food from various surfaces and from holes. They frequently hover and turn without translocation (Blake, 1976, 1977, 1978). Alternatively, these and others manoeuvre in open waters to position the body to suck small prey items from the water column.

Manoeuvrers are characterized by relatively large, flexible median and paired fins, and a short, deep body (Gosline, 1971; Alexander, 1967) so that MPF gaits are well developed. Station-holding gaits appear suppressed, and manoeuvrers avoid currents by hiding among structures (Keenleyside, 1979). The body and fins have a large surface area, which implies performance in BCF cruising and sprinting gaits should be suppressed. Similarly, the manoeuvrer body is often highly compressed, potentially reducing myotomal muscle so that fast-start gaits could be reduced.

The functional importance of the body morphology of manoeuvrers is based largely on the assumption that manoeuvrability is important because of the habitats such fish occupy. Experimental field data are rare, but supportive (Werner, 1986; Ehlinger & Wilson, 1988). Laboratory experimental data are equivocal. For example, angelfish are not more adept than fusiform fish in negotiating slits and tubes, illustrative of small gaps among habitat structures (see Webb, 1994a). On the other hand, angelfish are very agile (see below). Cruising and fast-start performance of manoeuvrers is only reduced compared with specialists for such BCF gaits (Beamish, 1978; Domenici & Blake, 1991; Webb, 1992).

As speed increases in MPF gaits, the number of active propulsors appears to decline, and different species tend to use preferentially one or a pair of fins (Blake, 1977; Webb, 1994a). This forms the basis for the MPF modes described by Breder (1926) (Figure 1). Variations probably correlate with life-style, but data are inadequate to draw firm conclusions. Thus bentho-pelagic amiiform fish use dorsal fin undulation. Balistiform fish, living among corals, emphasize dorsal and anal fin undulation, presumably for better balance of thrust forces in manoeuvring. Actinopterygian diodontiform species similarly use their pectoral fins. Electric fish need a stable long-body baseline, so that gymnotiform species undulate their ventral fins, keeping the body straight. Depressed benthic skates and rays use pectoral fin undulation in the rajiform mode. Oscillatory resistance-based dorsal and anal fins working as a pair are found in tetraodontiform swimmers, while labriform swimmers similarly oscillate pectoral fins. Lift-based oscillatory pectoral fins are used by actinopterygian embiotociform and elasmobranch mobuliform fish, while moliform fish use lift-based dorsal and anal fins. Such fish can cruise at speeds comparable to those of generalized BCF swimmers.

Cruiser/sprinters, accelerators and manoeuvrers tend to use the water column much of the time. Other fish are more closely tied to the substratum, living or moving through structures so dense that swimming may not be possible, or resting on the bottom in the presence of currents. The former are burrowers. In all animal groups, burrowing is associated with an elongate body form and reduction of projecting appendages. This body form is typical of anguilliform fish, which swim primarily by BCF gaits. However, the small body and fin depth does not entrain much water to generate high thrust, so that these fish do not cruise or sprint at high speeds (Beamish, 1978). Furthermore, the large number of body waves found within the body length is relatively inefficient (Lighthill, 1975). Therefore, swimming performance in the BCF gaits is low (Beamish, 1978). However, high swimming costs would make burst-and-coast swimming especially economical (Weihs, 1974). Some anguilliform fish make extensive migrations, albeit over long periods and at low speeds compared with cruiser/sprinter specialists. Burst-and-coast swimming might be important in these migrations.

Acceleration for a fast-start gait is also suppressed because of the small added mass of the body and fins of burrowers. Indeed, some anguilliform fish have reduced Mauthner cells, the principal pathway initiating an escape fast-start. When startled, they reflexly retract into cover instead of fleeing (Eaton *et al.*, 1977).

Many benthic and bentho-pelagic fish are exposed to substantial currents, for example in streams or tidal areas. For these, flow refuging is essential, but direct avoidance of current often is not possible. Specializations for station-holding gaits are associated with two major patterns of compromise, between lift that raises the body from the substratum causing it to be displaced downstream, and drag that tends to push a fish downstream (Arnold & Weihs, 1974). One form is somewhat depressed, typically in cottids (Figure 3), which has low lift but relatively high drag. Cottiform fish offset the drag with devices to increase friction, such as extensive fins capable of grasping the substratum (e.g. cottids, salmonids), creation of negative lift, especially with the pectoral fins (e.g. salmonids, catostomids), rough body surface (e.g. elasmobranchs, loricariids), or suction devices typical of the 'torrential' fauna living in high-flow streams (e.g. gobiesocidae) (see Webb, 1989, 1993, 1994a). High drag/low lift cottiform fish appear to suppress MPF and BCF sustainable gaits. This may reflect the prevalence of these fish in faster and more turbulent flow situations.

The second pattern used in station-holding is to minimize drag with a body form that is flattened parallel to the substratum and hence to the flow. This is characteristic of plaice (pleuronectiformes) and rays. This pleuronectiform shape is common in habitats where there may be insufficient surface roughness to increase friction to hold station. However, in order to maintain body volume, these 'flattened fish' expand the body normal to the flow, parallel to the substratum and hence have a large projected area and high lift. Various behaviours induce flow beneath the body to minimize lift (Arnold, 1969; Arnold & Weihs, 1978), and on a flat, smooth surface pleuronectiform fish are able to hold position at much higher current speeds than cottiform fish (Webb, 1989).

Ontogeny and size

The range of gaits expressed by fish increases during ontogeny (Figure 4), especially for marine species starting from a yolk-sac larva. Yolk-sac and first-feeding larvae move primarily in a viscous regime (Webb & Weihs, 1986; Fuiman & Webb, 1987; Fuiman, 1986; Batty, 1981). BCF-white muscle sprints and fast-start gaits are typical. The BCF-red muscle gait is probably poorly developed because larvae have only a single layer of red fibres (Batty, 1984). MPF gaits are suppressed because median fins develop late (Webb & Weihs, 1986).

At first feeding, larvae become elongate and can swim at high enough Reynolds numbers for inertial forces to be large. At this time the economical BCF burst-and-coast gait appears (Weihs, 1980). With further differentiation of red muscle fibre (Batty, 1984), lower speed BCF gaits are added (Figure 4).

As development proceeds, the paired fins differentiate, again when they can be large enough to exploit inertial forces (Webb & Weihs, 1986). Then MPF gaits are added. Finally, station-holding gaits may be added at metamorphosis for benthic species.

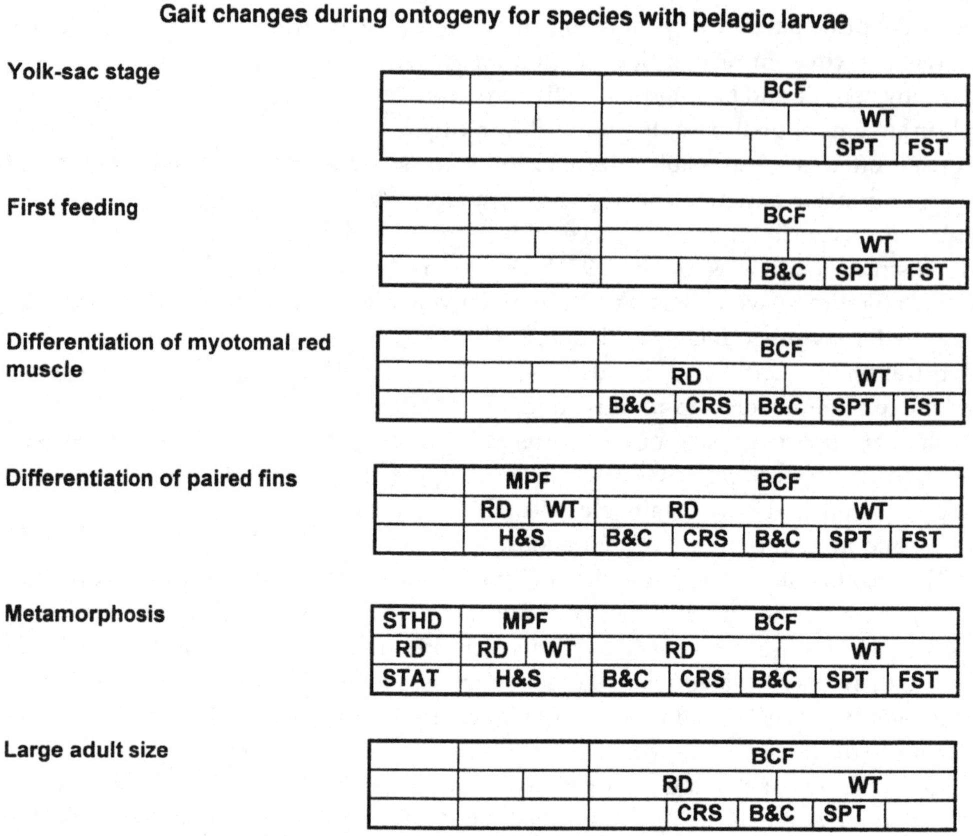

Gait changes during ontogeny for species with pelagic larvae

Yolk-sac stage

			BCF		
				WT	
				SPT	FST

First feeding

			BCF		
				WT	
			B&C	SPT	FST

Differentiation of myotomal red muscle

		BCF			
		RD		WT	
B&C	CRS	B&C	SPT	FST	

Differentiation of paired fins

MPF		BCF			
RD	WT	RD		WT	
H&S	B&C	CRS	B&C	SPT	FST

Metamorphosis

STHD	MPF	BCF			
RD	RD	WT	RD	WT	
STAT	H&S	B&C	CRS	B&C	SPT	FST

Large adult size

		BCF		
		RD	WT	
	CRS	B&C	SPT	

Figure 4. Gait changes associated with developmental stages. The key is the same as in Figures 1 & 2.

Where final adult size is large, post-metamorphic development may be associated with gait suppression (Figure 4). With increasing size, a mismatch grows between muscle force, $\propto \text{mass}^{0.67}$, and inertial resistance $\propto \text{mass}$ (Daniel & Webb, 1987). As a result, acceleration, and especially fast-start performance, is suppressed (Daniel & Webb, 1987; Webb & de Buffrénil, 1990).

Environmental factors and gait compression

Performance in various gaits may be affected by environmental conditions. Gaits important in fitness critical events supported by white muscle appear to be least affected (Hochachka & Somero, 1984; Webb, 1978b; Goolish, 1991; Webb & Zhang, 1994). This is similar to the situation in other ectotherms (Bennett, 1991).

Most research, however, has focused on how abiotic environmental elements affect performance in the BCF-red muscle gait (Beamish, 1978), using Fry's classification of environmental factors as a framework. This classification categorizes abiotic elements into factor classes according to common effects on production and distribution of metabolic energy (see Brett, 1979 for details). In general, temperature and pressure (controlling factors) determine the limits for energy delivery and hence maximum cruising speed. Other elements affect the supply and removal of metabolites (limiting factors), or shunt energy from locomotor to regulatory functions (masking factors) thereby reducing maximum cruising speed. Factors that alter physiological states (directive factors), for example for reproduction, have not been well studied but could facilitate or reduce performance.

Most abiotic environmental factors reduce metabolic scope and hence speed in BCF-red muscle gaits (Brett, 1979). The much smaller MPF-red muscle is probably affected only by controlling factors. Reduced speed in environmentally-sensitive gaits should result in earlier shifts to environmentally less sensitive white muscle gaits or burst-and-coast behaviour. Such gait compression has been observed only in response to low temperature (Rome *et al.*, 1990).

AREAS FOR FURTHER STUDY

Computational models

Biomechanics has been central to the development of ideas on swimming. For the first 60 years of this century, the dominant questions dealt with the magnitude of drag relative to energy availability (Alexander, 1983). In the late 1960s, attention shifted to problems of thrust production, particularly the application of elongated slender-body theory to many MPF and BCF gaits (Weihs, 1972, 1973b; Lighthill, 1975, 1990a,b,c; Wu, 1977; Daniel & Webb, 1987; Webb, 1988; Daniel *et al.*, 1992). Other hydromechanical models have been developed for thunniform lunate tail propulsion used by the fastest aquatic animals (Daniel *et al.*, 1992). In another direction, blade-element theory has been popular for studying fin motions in MPF gaits (see Blake, 1983a).

Swimming movements are very complex and all these approaches make simplifying assumptions. For example, elongated slender-body theory assumes amplitudes of swimming movements are small relative to the propulsive wavelength, and that thrust forces are reactive in origin. Yates (1983) points out that these assumptions may be met only marginally by swimming fish (see also Vlymen, 1974; Jordan, 1992).

Computational models are now being developed, made possible by the increasing speed of computers. These models seek to describe real fish motions and forces more closely, providing for more sophisticated and realistic evaluations of variation. As an example, Froude efficiency for BCF undulation is primarily determined from easily measured wave parameters in elongated slender-body theory. Efficiencies of similar magnitudes have been calculated for different types of swimmers, such as fish, snakes, tadpoles etc. (e.g. Wassersug, 1989) without consideration of the large effects of differences in morphology. Computation models clearly show variation in efficiency associated with different undulant patterns (W.W. Schultz, P. Zhou & P.W. Webb, unpublished data).

Manoeuvrability and agility

Questions of maximum performance continue to dominate studies of fish locomotion, focusing on speed and linear acceleration in BCF gaits. These performance limits often reflect a reserve capacity to deal with occasional but critical events such as migrations, spates in streams or predators. However, speed and linear acceleration are insufficient to capture the full range and limits of performance defined by velocity, acceleration, manoeuvrability and agility in three-dimensional space (Webb, 1994a).

Following the principles set out by Norberg & Rayner (1987), I define manoeuvrability as turning ability. It is measured as turning radius, and comparisons among species should be made at mechanically equivalent speeds. Agility is defined as the rate of turn. Both agility and manoeuvrability are measured in an orthogonal co-ordinate system fixed in the body. Rolling is a rotation about the anterio-posterior axis of motion. Yawing is rotation about the vertical axis normal to the axis of motion, and passing horizontally through the centre of inertia. Pitching rotations occur around an axis passing horizontally through the centre of inertia normal to the axis of motion.

Data on manoeuvrability and agility are few, and all are for yawing turns. Seahorses, *Hippocampus hudsonius* (Blake, 1976), *Lactotoria cornuta* and *Tetrasomun gibbosus* (Teleostei: Ostraciidae; Blake, 1977) and *Rhinecanthus aculeatus* and *Odonus niger* (Teleostei: Balistidae; Blake, 1978) can turn with zero-radius while hovering using MPF gaits. During BCF swimming, minimum turning radius is independent of swimming speed, supporting arguments by Howland (1974) that the magnitudes of centrifugal resistance and available turning force vary in the same way with speed (Webb, 1983). Minimum turning radius is also independent of acceleration (Webb, 1983), and, within a species, is a constant proportion of length (Webb, 1976). Minimum turning radii have been measured as $0.23L$ for yellowtail, *Seriola dorsalis* (Webb & Keyes, 1981), 0.17-$0.18L$ for rainbow trout, *Oncorhynchus mykiss* (Webb, 1976, 1983), 0.13 for dolphin, *Coryphaena hippurus* (Webb & Keyes, 1981), and 0.065 for angelfish, *Pterophyllum eimekei* (Domenici & Blake, 1991). A listing of fish by minimum turning radius is also a ranking from more thunniform fish, through sub-carangiform to more chaetodontiform fish, consistent with theoretical expectations (Webb, 1983).

Measurements of agility are also few. Maximum turning rates can be estimated from the relationships between manoeuvrability, size and speed for yawing turns. For example, using Wardle's (1975) method of estimating maximum sprint speed, u_{sprint}, from *in vitro* measures of muscle twitch times and stride length (distance travelled per tail beat, here taken as $0.75L$): $u_{sprint} \approx 40 \ L^{0.6}$. Since turning radius is proportional to length, agility decreases with increasing size. Consequently, smaller fish are both more manoeuvrable and more agile.

I know of no data on pitching and rolling manoeuvrability and agility, rotations for which fin trimmers and body shape confer high stability and low manoeuvrability (Alexander, 1967; Gosline, 1971; Aleyev, 1977). Body flexibility is also low in all but lateral movements. As a result, large turning forces are difficult to generate in planes other than yawing. Fish that attack prey silhouetted at the water surface may have greater agility and manoeuvrability in pitching, and negatively buoyant fish that must bank while turning (Weihs, 1981) may be more agile in rolling.

Stability

Stability and manoeuvrability/agility are limits of a continuum of posture control problems in locomotion. In general, a stable body remains on a given trajectory, returning to that trajectory if perturbed, thereby preventing manoeuvring. Conversely, highly manoeuvrable bodies tend to be unstable.

Problems and mechanisms involved in regulating stability have received relatively little attention, with the exception of hydrostatic equilibrium and trimming (Alexander, 1967; Aleyev, 1977; Weihs, 1989). The paucity of data reflects the traditional research emphasis on high levels of performance where stability is not an obvious problem. Forces generated by propulsors and acting on the fins are proportional to u^2, and hence are large relative to destabilizing forces at high swimming speeds (Aleyev, 1977; Marchaj, 1988).

Controlling posture may be a major problem during hovering and slow swimming because thrust and stabilizing forces quickly become small compared to the inertia which must be manoeuvred (Marchaj, 1988). Numerical modelling of pressure fields suggests coasting is intrinsically unstable (W.W. Schultz, P. Zhou and P.W. Webb, unpublished data). The use of fins for stability is believed to explain higher than expected occurrences of drag of coasting fish (see Webb, 1975a; Blake, 1983a).

Sources of instability are many. Fish are unstable in hydrostatic equilibrium because the centre of mass is usually located above the centre of buoyancy, while pressure-sensitive gas inclusions confer neutral buoyancy at only one depth (Gee, 1983; Alexander, 1989a). Extrinsic destabilizing forces arise from asymmetries in the flow about the body and fins, currents induced by other animals, the flow around habitat structures, and wind-induced flows. All become large relative to the momentum of a fish hovering or swimming slowly.

Stability control may loom large in determining routine energy expenditure, a major feature in energy budgets affecting that available for reproduction (Brett & Groves, 1979). Mechanical analyses suggest the cost of hovering is very much larger even than that of slow swimming (Blake, 1979). Stability problems are probably especially prevalent in routine swimming, which is increasingly found to involve speeds very much lower than a traditional one length per second 'rule-of-thumb'. Energy costs may be up to an order of magnitude higher than previously thought (Webb, 1991; Block et al., 1992; Boisclair & Tang, 1993; Krohn & Boisclair, 1994).

The stability problems confronting fish swimming at low speeds have long been recognized for negatively buoyant pelagic cruiser/sprinter specialists (Harris, 1936, 1937, 1938; Magnuson, 1978). The body weight of these fish is supported by hydrodynamic lift generated by the body, caudal peduncle, paired fins and caudal fin. High aspect-ratio pectoral fins (see Figure 4) generate lift and a positive pitching moment ahead of the centre of mass. This is balanced by lift and a negative pitching moment aft. The caudal peduncle of thunnids generates the negative pitching moment, but the asymmetrical tail serves this function in selachians (Harris, 1936, 1937, 1938; Alexander, 1965, 1990; Magnuson, 1978; Thomson, 1976; McGowan, 1992). Variable pectoral fin area and angle of attack trim lift forces and moments (Bone & Marshall, 1982; Weihs, 1989).

Downward orientation of caudal fin thrust can also occur (Thomson, 1976; Thomson & Simanek, 1977), and this involves whole-body tilting as speed decreases (He & Wardle, 1986). Neutrally buoyant fish do not share the same specific problem as negatively buoyant fish of stabilizing body position in the water column, but must still stabilize their posture. As with negatively buoyant fish, neutrally buoyant fish may increasingly tilt as speed decreases (Webb, 1994b). For both negatively and neutrally buoyant fish, tilting increases drag, requiring higher thrust. The increased thrust can be viewed as providing a reserve better matched to destabilizing forces. Presumably stabilized, slow swimming with elevated drag is less costly than unstable motions.

Increased thrust at low swimming speeds, presumably associated with stability control, appears to be the general rule for swimmers. It is well known that the relationship between propulsor kinematics and swimming speed does not pass through the origin (Videler & Wardle, 1991) such that thrust (= drag) coefficients for fish swimming slowly (at low Reynolds numbers) are much higher than expected (see Webb, 1975a, 1993, 1994a; Blake, 1983a; Videler & Wardle, 1991).

Although manoeuvrability and stability are the antitheses of each other, fish are both manoeuvrable and stable. This appears possible only in the plane of lateral body bending where yawing stability is a dynamic balance between the side force of the tail resisted by the virtual mass of the anterior body (Lighthill, 1975; Webb, 1992). This dynamic situation can easily be destabilized by asymmetrical motions akin to fast starts increasing the force of the posterior body while steering with the head (Weihs, 1972, 1973b). This results in small radius turns at high rates.

CONCLUSIONS

There are numerous complementary approaches to the study of fish locomotion. The approach I have used here is one which I suggest has merit for a number of reasons. First, it escapes from the focus on performance maxima, redirecting attention to general problems of swimming faced by all fish. Second, it provides a framework that incorporates other approaches such as biomechanics and natural history (see Werner, 1986; Winemiller, 1991; Daniel et al., 1992). For example, differences in MPF swimming modes are seen as variants that fulfil a generally similar ecological function (slow swimming). Third, the approach provides a simple way to summarize and explore temporal changes in both evolutionary and ontogenetic time scales. Finally, environmental effects are readily incorporated. In general, an intraspecific starting point based on gaits for performance range fractionation appears to unify many different biological and mechanical aspects of locomotion. It does so in a way that highlights large areas deserving of study.

This work was supported by National Science Foundation grant no. DCB9017817.

Chapter 5

Swimming physiology of pelagic fishes

J.B. GRAHAM*, H. DEWAR*, N.C. LAI*, K.E. KORSMEYER*,
P.A. FIELDS*, T. KNOWER*, R.E. SHADWICK*, R. SHABETAI[†],
AND R.W. BRILL[‡]

*Center for Marine Biotechnology and Biomedicine and the Marine Biology Research
Division, Scripps Institution of Oceanography, University of California, San Diego,
La Jolla, CA 92093, USA. [†]Department of Medicine, UCSD and Veterans
Administration Medical Center, San Diego, CA 92161, USA. [‡]National Marine
Fisheries Service, Honolulu, HI 96822, USA

A large, portable water tunnel has permitted insight into the swimming physiology of
sharks and tuna. Tunnel use has demonstrated the capacity of sharks to sustain swimming
speeds of over one body length per second. This level of aerobic performance for sharks
and the measured rates of oxygen consumption during swimming both exceed previous
estimates for this group, and are similar to values reported for bony fishes. By permitting
studies with stably swimming, instrumented sharks, the tunnel has enabled experimental
analyses demonstrating the *vis a tergo* cardiac filling mechanism and the role of the
pericardioperitoneal canal in facilitating rapid changes in cardiac stroke volume. For
tunas, the tunnel has permitted long-term and replicate studies of metabolic rate and
factors affecting it (temperature, body size, velocity), of heat transfer, and of swimming
kinematics. With the incorporation of anaesthesia and surgical techniques it is now
possible to investigate tuna cardiorespiratory responses to changes in temperature, oxygen, and swimming velocity, and to carry out biomechanical studies that include the
timing of electromyograms at sites along the body, and measurement of muscle force
generation in the lateral tail tendons during swimming.

INTRODUCTION

A major focus of our laboratory has been the development of a large volume,
portable, high-speed water tunnel respirometer to study the swimming performance
and physiology of fishes such as tunas (Figure 1) and many sharks. Prior to the

Maddock, L., Bone, Q. & Rayner, J.M.V. (ed.). *Mechanics and Physiology of Animal Swimming.*
© 1994. Cambridge University Press.

Figure 1. Yellowfin tuna swimming in the water tunnel. The overhead 45° mirror permits
simultaneous top and side views. Grid squares are 10x10 cm.

completion of this unit there had been no studies of shark swimming capacity (Graham
et al., 1990) and only a few studies of tuna swimming performance (reviewed in Dewar
& Graham, in press a).

A recent paper (Graham *et al.*, 1990) has fully detailed and diagrammed the SIO
tunnel structure. The unit's basic design follows that of a Brett respirometer (Brett,
1965). It is comprised of an oval loop of 46-cm (18-inch) diameter pvc pipe (upgraded
from 30 cm for the tuna work) containing the following specialized, in-series sections: a
propeller housing containing a 46-cm, 12-pitch propeller, a diffuser-contraction (DC)
section that minimizes turbulence, and a 100x51x42 cm (lxwxh) working section that
holds the experimental fish. Flow straighteners located between the DC and the working section help to establish microturbulent flow. The tunnel is supported on a steel I-
beam frame that is 7·3 m long and just over 3 m wide. The unit's mass (without water)
is 4000 kg. The propeller is powered by a 40 hp electric motor (440 V) operated at
variable speed by an AC controller. Water velocity can be adjusted from 0 to over 3 m
s^{-1}. The system's volume is 3000 litres which includes a separate filtration and chilling

loop supplied by a 1·4 hp centrifugal pump. The working section contains a removable, clear lucite lid and a clear lucite outer wall. A 45° mirror over the working section permits simultaneous top and side views of the swimming fish (Figure 1). Using this system it is now possible to conduct physiological, energetic, and kinematic studies on sharks and tunas after they have recovered from handling stress, and in an experimental setting in which ambient conditions are regulated and where swimming speed itself is a controlled experimental variable.

The major objective of our work has been to compare the manner in which basic fish baupläne and supporting cardiorespiratory and neuromuscular functions have been integrated for optimal swimming. In view of the recent analysis (Alexander, 1991) suggesting little or no apparent adaptive advantage for body form in relation to either maximum swimming velocity or swimming energetics, broadly comparative studies, contrasting species as diverse as sharks and tunas, are needed to define key morphological, kinematic, physiological, and biochemical specializations for optimal swimming. The aims of this paper are to summarize the variety of data obtained in recent and on-going water tunnel studies, and to evaluate these findings in the context of basic questions about fish swimming performance.

SHARKS

Past investigations of shark swimming performance, physiology, and metabolism were limited by the size of available water tunnels and the difficulties of maintaining sharks in captivity (Bone, 1988). As a result, generalizations about shark swimming physiology are based on studies in which activity was not controlled and small specimens of less active species were used (Graham et al., 1990).

Protocols were developed to investigate the physiology and energetics of swimming sharks. This work focused on measuring the effects of swimming on the haemodynamics and oxygen transport of the California leopard shark (*Triakis semifasciata* Girard) (Lai et al., 1989, 1990a,b). Studies were also done on ventilatory pattern and resting *vs* swimming metabolic rates, with data being obtained for *Triakis* and the lemon shark (*Negaprion brevirostris* Poey) as well as, during sea-going studies, the mako shark (*Isurus oxyrinchus* Rafinesque) (Scharold et al., 1989; Graham et al., 1990).

Swimming and shark heart function

Controlled swimming experiments enabled testing, for the first time, of two important hypotheses about elasmobranch heart function. The first is the long-held idea that a negative (*i.e.* subambient) pericardial pressure is an essential requirement for elasmobranch heart function. Since Schoenlien's (1894) classic observation, the obligatory dependence of the elasmobranch heart on a negative (*vis a fronte*) filling pressure has been a cornerstone of comparative cardiology (Satchell, 1991; Farrell & Jones, 1992). Recent experiments, however, showed that transmural heart (*i.e.* pericardial) pressures were not always negative, and that these could be an artefact of the loss of pericardial fluid volume during specimen handling (Shabetai et al., 1985; Abel et al., 1986; Lai et al.,

1989). We tested the negative pressure hypothesis by comparing vascular and pericardial pressures in resting and swimming sharks. The second hypothesis concerns the role of the pericardioperitoneal canal (ppc). This structure, which is found in elasmobranchs and primitive bony fishes, is a one-way conduit between the pericardium and the peritoneum. There had been no theories about the function of this structure until our initial experiments suggested it is a pressure-relieving mechanism allowing rapid increases in diastolic heart volume (Shabetai *et al.*, 1985).

Specimens of *Triakis* were instrumented with vascular, ventricular, and pericardial catheters, and with a ventral aortic flow meter in order to monitor blood and ventricular pressures, pericardial pressure and volume, and cardiac stroke volume. These variables were then compared in resting fish and during swimming at a controlled velocity (U).

To examine the importance of negative pericardial pressure in cardiac filling, simultaneous measurements of pressure in the central venous circulation (cardinal sinus) and pericardium were made in resting and swimming fish. Assuming that a negative-pericardial-pressure-induced suction was the agent of cardiac filling, then a zero pressure gradient would be expected between the heart and cardinal sinus. Sinus pressure could also be expected to oscillate in phase with the cardiac cycle and become negative during ventricular diastole. These, however, were not the results of our studies (Lai *et al.*, 1989). Resting fish most commonly had a positive mean pericardial pressure. Also, cardinal sinus pressure exceeded that in the pericardium, and cardiac-induced fluctuations in cardinal sinus venous pressure were considerably less than in the pericardium. Simultaneous pressure records for the ventricle and pericardium further showed that ventricular diastolic transmural pressure was consistently positive. This means that *vis a fronte* (suction) filling was not occurring in resting sharks and that the dominant force for cardiac filling is a positive central venous pressure (Lai *et al.*, 1989).

The effect of swimming was to increase both sinus and pericardial pressure but also to accentuate the differences between them; pericardial pressure increased by approximately 0·1 kPa whereas ventricular diastolic pressure rose by several kPa. Swimming therefore generates a substantial increase in ventricular diastolic transmural pressure, which means that positive venous pressure can be the only mechanism to account for cardiac filling in *Triakis*, and most likely all elasmobranchs.

Swimming experiments also demonstrated that the ppc's function as an overflow relief valve which allows an increased cardiac diastolic volume in support of increased metabolic demand. As seen in Figure 2, swimming elevates pericardial pressure sufficiently to eject pericardial fluid from the ppc, and there is good agreement between fluid ejection volume and the increase in average cardiac stroke volume during swimming (Lai *et al.*, 1989).

Swimming performance and respiration

The conclusions of most previous studies were that sharks have lower metabolic rates than teleosts and little capacity for sustained aerobic swimming (Piiper *et al.*, 1977; Brett & Blackburn, 1978; Bushnell *et al.*, 1982; reviewed by Graham *et al.*, 1990). We therefore conducted a comprehensive set of tests of the aerobic capacity of *Triakis*, measuring

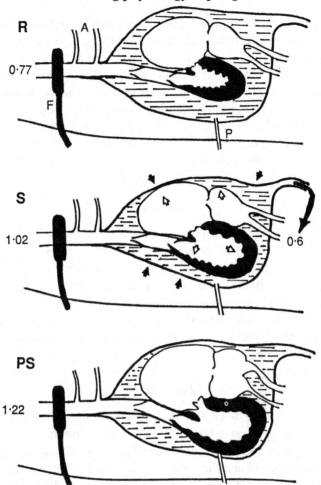

Figure 2. Diagram of leopard shark heart, ventral aortic, and pericardial structure illustrating the dynamic effects of swimming. (R) Conditions during rest showing relatively large ratio of pericardial to heart volume filled by pericardial fluid. Flowmeter (F) and catheter access for pericardial pressure and volume (P) are shown. (A) shows two afferent branchial arches emanating from the ventral aorta, and resting cardiac output is 0·77 ml kg^{-1}. (S) Effects of swimming are compression of pericardium by contiguous body muscles (Figure 1) and an increased cardiac volume resulting from elevated venous return. These cause a 0·6 ml kg^{-1} pericardial fluid loss (arrow) and increase cardiac output to 1·02 ml kg^{-1}. (PS) Post-swimming recovery is characterized by further increased cardiac stroke volume resulting from reduced compressive forces and fluid loss.

critical swimming speeds (U_{crit}, an index of aerobically sustainable swimming speed) and the effects of velocity on oxygen consumption (Graham et al., 1990) and on blood gas parameters (Lai et al., 1990a). Endurance tests show that Triakis can sustain swimming speeds of over one body length per second (L s^{-1}) and that its critical swimming velocity varies inversely with body size, from about 1·6 L s^{-1} in 30-50 cm sharks to 0·6 L s^{-1} in 120 cm sharks. These critical speeds are only slightly less than those for comparably-sized sockeye salmon (Graham et al., 1990).

In Triakis, sustained swimming (i.e. at speeds $<U_{crit}$) increases cardiac stroke volume

by over 30% (Figure 2), raises mean heart rate by about 10%, and elevates ventral aortic pressure; it also reduces both venous pH and blood PO_2 (*i.e.* sustained aerobic activity decreases the venous oxygen reserve to a new steady level). Also occurring are increases in both venous PCO_2 and circulating lactate. While these tests were done at sustainable speeds, the rise in circulating lactate indicated that some anaerobic muscle units had also increased their activity. During rest periods of up to one hour, lactate levels dropped but remained elevated and the fish remained slightly acidotic. Neither oxygenated nor deoxygenated blood pH were affected by CO_2 equilibration, suggesting the absence of a Haldane effect (Lai *et al.*, 1990a).

TUNAS

In 1990 the water tunnel was transported to the Kewalo Basin Research Facility of the US National Marine Fisheries Service in Honolulu, Hawaii, to carry out investigations of tuna swimming physiology and biomechanics. In addition to their ecological and commercial importance, tunas continue to hold the interest, if not fascination, of biologists because of their numerous specializations for continuous, efficient locomotion. From a highly streamlined body shape to a heightened capacity for aerobic performance augmented by endothermy, the tunas embody the quintessential adaptive state for continuous swimming.

Energetics

A major objective was to determine the metabolic costs of tuna swimming and compare these with other cruise-adapted teleosts such as the sockeye salmon. Studies with immobilized tunas show that the physiological and biochemical adaptations underlying their heightened aerobic capacity impose a high standard metabolic cost (Brill, 1987; Brill & Bushnell, 1991). However, Boggs & Kitchell (1991) wrote, "there is no strong evidence that the increase in metabolic costs with swimming speed is less steep in tuna than in active, cold-blooded fishes". We therefore tested the hypothesis that the numerous morphological specializations of the tuna bauplan do translate into a lower rate of increase in swimming costs with velocity. The water tunnel enabled long-term studies of tuna metabolism at controlled U and ambient temperature (T_a) and allowed estimation of the underlying standard metabolic maintenance costs (Dewar, 1992; Dewar & Graham, in press a).

Swimming oxygen consumption rates ($\dot{V}O_2$) were determined for three tuna species (yellowfin, *Thunnus albacares* Bonnaterre; kawakawa, *Euthynnus affinis* Cantor; and skipjack, *Katsuwonus pelamis* L.) during stable swimming over a range of speeds (17-150 cm s^{-1}) and T_as (18-30°C). The specimens studied ranged in body size from 30-50 cm fork length (FL) (mass 0·5-3·5 kg). Major effort was focused on the yellowfin because it was generally more available, came in a greater size range, did well in captivity, and readily adapted to the tunnel. Metabolic experiments covered long periods after handling (the maximum test record was 31 h; this test was ended due to our fatigue and not that of the fish) and replicate experiments on subsequent days were carried out on several specimens. We determined the relationship between U and $\overset{\bullet}{V}O_2$ for yellowfin and skipjack and, by extrapolating to $U=0$, estimated the standard $\overset{\bullet}{V}O_2(S\overset{\bullet}{V}O_2)$ of differently-sized groups (Dewar & Graham, in press a).

Compared to the sockeye salmon, the yellowfin tuna has much higher standard metabolic costs (Figure 3). The $S\overset{\bullet}{V}O_2$ for a 2·2 kg yellowfin is 257 mg O_2 kg^{-1} h^{-1} (24°C) whereas that for a 1·4 kg sockeye is 44 mg O_2 kg^{-1} h^{-1} (15°C). This disparity is not attributable to either body mass or temperature differences. To illustrate this point, a doubling of the sockeye's $\overset{\bullet}{V}O_2$ to correct for temperature (*i.e.* assuming a Q_{10} of 2 from 15 to 24°C) still results in a value much below that of the yellowfin which, because it has a greater mass, should be expected to have a lower mass-specific $\overset{\bullet}{V}O_2$ (*i.e.* mass-specific $\overset{\bullet}{V}O_2$ should be μmass$^{-0·25}$). Thus the yellowfin's higher metabolic rate reflects a suite of underlying physiological and metabolic adaptations supporting the heightened aerobic capacity (Brill, 1987; Brill & Bushnell, 1991).

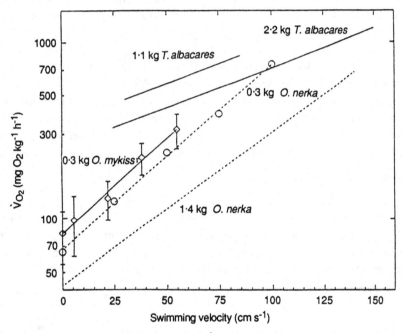

Figure 3. Comparison of salmonid and scombrid $\overset{\bullet}{V}O_2$ data plotted as a function of U. Data for 0·3 kg (circles, dashed line) and 1·4 kg (dashed line) sockeye salmon (*Oncorhynchus nerka*, Brett, 1965) at 15°C. Also shown are the mean $\overset{\bullet}{V}O_2$ ±SD (diamonds, solid line) for a 0·3 kg rainbow trout (*O. mykiss*, Bushnell *et al.*, 1984) at 15°C. The solid lines show linear regressions for 42 cm FL (1·1 kg) and 51 cm FL (2·2 kg) yellowfin at 24°C.

Our work has also verified the hypothesis that the specialized tuna bauplan increases swimming efficiency (Dewar & Graham, in press a). Figure 3 shows the rate of increase in $\overset{\bullet}{V}O_2$ with U is much less for the yellowfin than for the sockeye. The $\overset{\bullet}{V}O_2$ of a 2·2 kg yellowfin is about 300 mg O_2 kg^{-1} h^{-1} at a U of 25 cm s^{-1} and becomes 1100 mg O_2 kg^{-1} h^{-1} at 140 cm s^{-1} (a factor of 3·7). By contrast, the sockeye's rate increases by a factor of 10; from 70 mg O_2 kg^{-1} h^{-1} at 25 cm s^{-1} to 700 mg O_2 kg^{-1} h^{-1} at 140 cm s^{-1}.

Thermoregulation

An extensive literature surrounds the capability of tunas to regulate body temperature and rates of heat transfer, and the experimental approaches have varied from telemetry tracking of tuna body temperature (T_b) to laboratory tests of the effects of ambient temperature (T_a) on swimming speed and T_b (Dizon & Brill, 1979; Graham, 1983; Holland *et al.*, 1990, 1992). Using steadily swimming yellowfin and kawakawa fitted with a red-muscle thermocouple, we examined tuna capacity to modulate heat gain and loss in response to changes in U and T_a (Dewar *et al.*, 1991, in press).

The hypothesis tested was that these fish had the intrinsic capacity to regulate their rates of heat gain or heat loss and thus could 'physiologically thermoregulate'. Two large capacity (3000 l) sea-water reserve tanks, one containing chilled (0°C) and the other heated (45°C) water were incorporated into these tests. By pumping water from these tanks into the tunnel it was possible to impose square-wave changes in T_a on a steadily swimming fish in a fashion mimicking thermal changes encountered during vertical foraging sojourns. For example, a yellowfin in surface waters (25°C) might rapidly swim down into 18°C water in as little as 60-180 s, stay at that depth for a short period, and then return to surface waters (Holland *et al.*, 1990; 1992).

The protocol exposed fish to three up-down cycles of T_a change. This is illustrated in Figure 4 which shows T_a - T_b relationships for yellowfin no. 8 before and following each cycle change. By monitoring T_b during controlled changes in T_a it was possible to calculate the thermal rate coefficient (k, an index of the internal and external conductive properties contributing to heat transfer) in relation to other parameters in the heat transfer equation:

$$dT_b/dt = k(T_e - T_b) + Hp \tag{1}$$

where dT_b/dt is the rate of change of the initial T_b, T_e is the steady-state body temperature at the new T_a, and Hp is the extent to which the red muscle is warmed by its metabolic heat production. Because red muscle powers sustained swimming, Hp, although not directly measured, was regulated in these studies by our ability to control U.

Experiments show that the yellowfin modulates heat transfer; k is dependent on both

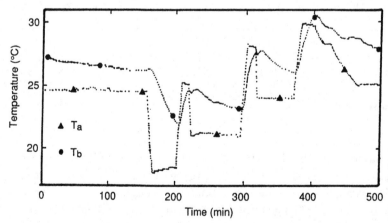

Figure 4. Effects of abrupt changes in tunnel water temperature (T_a) on the body temperature (T_b) of a 48 cm (1·8 kg) yellowfin tuna swimming at a speed of 0·8 L s⁻¹ during the entire test.

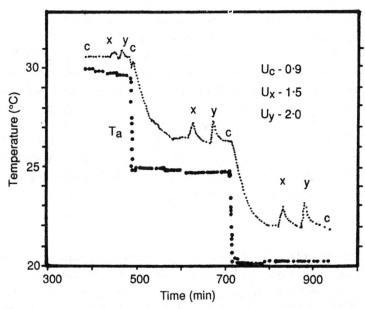

Figure 5. Effects of T_a and swimming velocity (U) on the T_b of a yellowfin tuna. Except during velocity tests x and y, fish swam continuously at control (c) speed.

T_a and the direction of the thermal gradient (*i.e.* fish coming up from depth may have a T_b less than surface T_a). Modulation of k in response to T_a was further demonstrated during tests in which U was varied; the elevation of T_b in response to equal increases in U was 3-4 times less at 30°C than at 25 and 20°C (Figure 5). There can be little doubt that endothermy augments tuna biology in a number of significant ways (Graham, 1983; Block *et al.*, 1993). However, just as important as heat retention to a tuna is the ability to regulate heat flux, which has utility in preventing overheating during intense activity, in retarding heat loss during a descent into cool water, and permitting heat gain upon returning to the warm surface waters (Dewar *et al.*, 1991, in press).

Kinematics

Several kinematic parameters have been measured for yellowfin tuna at different velocities in order to test the hypothesis that extreme differences in the tuna bauplan relative to other teleosts (*e.g.* a fusiform body shape, rigid vertebral column, insertion of myotomes over a greater number of vertebrae, paired and median fin shape and position, and red muscle amount and position) would be expressed in basic differences in key kinematic parameters (Dewar & Graham, in press b). Prior to this work there had been only a few studies of tuna swimming kinematics, primarily because of the limited opportunities to make detailed observations on fish swimming steadily at a known U (Fierstine & Walters, 1968; Magnuson, 1978). The investigated parameters included tail-beat frequency, caudal amplitude, yaw, stride length, the propulsive wavelength and propulsive wave velocity, and pectoral fin sweep-back angle. Among these, only two, the propulsive wavelength (also termed the bending wavelength) and, correspondingly, propulsive wavelength velocity, exceeded the values typical for cruise-adapted teleosts such as salmonids (Dewar & Graham, in press b).

Propulsive wavelength is the key kinematic parameter distinguishing yellowfin and probably all tunas from other teleosts. In yellowfin, propulsive wavelength exceeds fork length by a factor between 1·23-1·29, which is 30-60% longer than in salmonids. The morphological basis for the extended tuna propulsive wavelength lies in numerous modifications contributing to overall body stiffness listed above. This stiffness focuses axial muscle force on the caudal fin rather than on body segments (Fierstine & Walters, 1968; Magnuson, 1978). Another new observation is that swimming yellowfin have a distinctive yaw. Previous descriptions imply that yaw does not occur in 'the thunniform swimming mode' (Lindsey, 1978; Block *et al.*, 1993).

Cardiovascular

The central objective of this work has been to define cardiovascular parameters that set tunas apart from other cruise-adapted teleosts. Tuna swimming studies are now being conducted on fish fitted with percutaneous ECG (electrocardiogram) wires, with Doppler flow crystals mounted over the ventral aorta, and with both ventral and dorsal aortic cannulae to permit measures of: ventral and dorsal aortic blood pressure, respiratory gas tensions and pH, the levels of blood haemoglobin, lactate, and glucose and the amounts and types of circulating catecholamines. While this is not the first work to monitor cardiovascular variables in live tunas (Bushnell & Brill, 1992), our studies are the first to record these parameters for strongly swimming fish, several hours after anaesthesia recovery and in a setting where U, T_a, and water oxygen content are all controlled.

Current studies focus on two recent hypotheses about specializations in tuna cardiovascular function. Most fishes (including sharks, see above and Figure 2) elevate cardiac output during swimming by increasing both cardiac stroke volume and heart rate (Kiceniuk & Jones, 1977; Farrell & Jones, 1992). Farrell (1991) has suggested that, unlike all other fishes, and in a manner similar to mammals, tunas modulate cardiac output predominantly through changes in heart rate rather than stroke volume. Another hypothesis, offered by Brill & Bushnell (1991), is that because of the tuna's high blood oxygen capacity, cardiac output does not have to increase as much as in other fishes to meet the metabolic costs of increased speed. If either of these ideas is correct, it would mean that tuna hearts function differently from those of virtually all other fishes. In actuality there are very few data supporting these two hypotheses. Published values for tuna heart rate vary from as low as 60 to over 200 beats min^{-1} (bpm) and neither cardiac output nor arterial-venous oxygen differences have been measured in relation to U (Brill & Bushnell, 1991; Farrell, 1991).

Our water-tunnel studies have investigated heart rate and stroke volume during steady swimming in an experimental design analagous to that used to define U_{crit}. While studies are still in progress, data obtained thus far show that at velocities up to 3 FL s^{-1} heart rates have rarely exceeded 120 bpm (Korsmeyer *et al.*, 1992). Figure 6 shows a 9-h experiment measuring the relationships between U and both cardiac output and heart rate for a yellowfin tuna. Following an initial 2-h recovery period, U was increased stepwise, over a 5-h period, from 1·1 to 2·4 FL s^{-1} (54-118 cm s^{-1}). Heart rate ranged from 51 to 78 bpm (+53%) and relative cardiac output increased by a factor of about 1·4. At Us up to 2 FL s^{-1} heart rate remained fairly constant, and even declined slightly, and increases in cardiac output occurred solely by means of larger stroke volume. At 2·4 FL s^{-1}, a speed the fish could only maintain for about 15 min, heart rate increased substantially and relative stroke volume declined. These results clearly demonstrate that changes in both stroke volume and heart rate are important in providing increased oxygen delivery at higher U and thus emphasize basic similarities in the mechanisms for elevating cardiac output used by the yellowfin tuna and other fishes. None of these results supports conjectures (described above) about unique functional properties of the tuna heart.

Biomechanics

With the objectives of describing the mechanics of thunniform swimming, our group is carrying out experiments on swimming tuna fitted with a series of electromyogram (EMG) wires and a tail-tendon force transducer. These studies, which feature the first ever *in vivo* measurements of tendon forces in a swimming fish, are carried out at different velocities and T_as and all data are acquired simultaneously with dorsal (*i.e.* using the 45° mirror, Figure 1) video kinematic records (Knower *et al.*, 1993a, b).

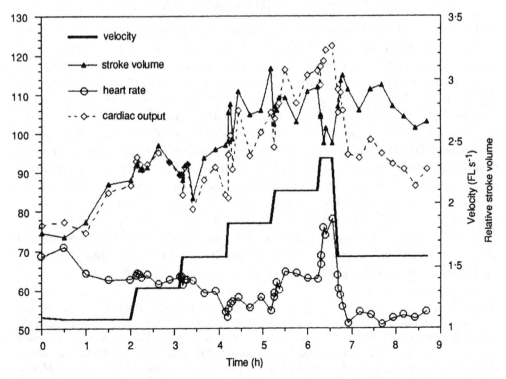

Figure 6. Changes in relative cardiac output and stroke volume and heart rate with velocity in a 48·5 cm (1·8 kg) yellowfin tuna at 25°C.

EMG wires had been implanted at several locations along the length of the superficial and core red (slow, aerobic) musculature of yellowfin and skipjack tunas, and muscle potentials were measured while fish swam at controlled speeds. As in other studies, we found that the onset of red muscle activation proceeded sequentially from the rostral to the caudal end of the body and burst durations became progressively shorter caudally. All EMG activity along the body turned off simultaneously. This muscle activation pattern is similar to that which has been described for mackerel and saithe by Wardle & Videler (1993) and for scup by Rome *et al.* (1993). However, unlike the general pattern for anguilliform and some sub-carangiform swimmers (Williams *et al.*, 1989; Blight, 1977) there is no overlap in EMG activity between the two sides of the body. Preliminary data further suggest that all the red fibres within a given cross-sectional plane, extending from the superficial to core layers, are synchronously activated and deactivated.

Simultaneous recordings of EMG and tendon force (using a stainless steel 'E'-shaped buckle force transducer placed on the great lateral tail tendons) show that all muscle activation on a side turns off just before peak force is registered in the tendons on that side, which occurs at maximum excursion of the caudal peduncle. Over a range of red-muscle-powered swimming speeds up to as fast as 102 cm s^{-1} in yellowfin and 133 cm s^{-1} in skipjack, tendon forces were found to be independent of speed and reached a maximum of 3·4 N in yellowfin and 3·8 N in skipjack.

SUMMARY AND CONCLUSIONS

The SIO Water Tunnel has permitted new insight into the energetics, physiology, and biomechanics of fish species that formerly could not be studied owing to their size and requirement for constant swimming. Also, the tunnel's portability, that is the ability to take it to sea to study mako sharks or to Hawaii for tuna work, is of fundamental importance. Tunnel experiments have successfully documented a far greater aerobic capacity for sharks than had been previously supposed. Also, shark swimming experiments were crucial to demonstrating the primacy of central venous pressure as the mechanism of cardiac filling, and in establishing ppc function. The tunnel has provided the means to maintain tuna at controlled swimming speed for long periods and then to regulate experimentally variables such as U, T_a, and oxygen. Sustaining stably swimming and viable specimens for long periods was the critical step needed to carry out studies of tuna metabolism, kinematics, and heat transfer. The incorporation of anaesthesia and surgical techniques has given added dimension to the tunnel's experimental utility with tunas and has paved the way for studies of thunniform biomechanics and cardiorespiratory function in relation to velocity and ambient conditions.

This work was supported by NSF grants (OCE 89-15927 and OCE 91-03739), by the Medical Research Service of the San Diego Veterans Administration Medical Center, by grants from the UCSD Academic Senate, by the M. Mazzerini Charitable Trust, Academic Rewards for College Scientist Foundation, Inc., the William H. and Mattie Wattis Harris Foundation, and by the Marine Biology Research Division, SIO, UCSD. Parts of the work were also facilitated by the National Marine Fisheries Service, Honolulu Laboratory. Ship time was provided by the University of California and we thank the captain and crew of RV 'Robert Gordon Sproul' for assistance and support.

Chapter 6

The mechanical design of the fish muscular system

LAWRENCE C. ROME

Department of Biology, University of Pennsylvania, Philadelphia, PA 19104, USA
and Marine Biological Laboratories, Woods Hole, MA 02543, USA

There are many components that make up the muscular system and some show more than 1000-fold variation in nature. Why is there all this variation and by what rules does evolution set the value for a component in a particular muscular system? The results from my laboratory show that there are three simple rules or design goals by which evolution sets the values for these components. First, fibre gear ratio (Δ body movement/Δ sarcomere length) and myofilament lengths are adjusted such that the muscle fibres being used always operate at optimal myofilament overlap where near-maximal force is generated. Second, gear ratio and the maximum velocity of shortening (V_{max}) are adjusted so that muscle fibres being used operate at the V/V_{max} (where V is the velocity at which the muscle is shortening) where maximum power is generated and where experiments on isolated muscle suggest that efficiency is nearly optimal. Third, the rate of activation and the rate of relaxation are adjusted to be relatively slow, which may permit mechanical power to be produced with optimal efficiency. Thus the way animals produce both slow, low frequency movements and fast, high frequency movements, is by recruiting different fibre types with the appropriate gear ratio, V_{max} and kinetics of activation and relaxation.

INTRODUCTION

One of the most fascinating areas of physiology is the study of how the parameters of a given system are fine-tuned to provide optimal performance under a variety of conditions. This chapter examines how the mechanical properties of the muscular system of fish are designed to power swimming. The fish muscular system provides an exceptional model to examine the general principles of physiological design. First, the performance at the level of the whole animal (swimming) can be easily studied and is intuitively important (it makes sense for fish to swim rapidly and/or efficiently).

Maddock, L., Bone, Q. & Rayner, J.M.V. (ed.). *Mechanics and Physiology of Animal Swimming.*
© 1994. Cambridge University Press.

Second, because of the regular geometry of the muscle, integration from the level of the cross-bridge to whole animal movements is relatively straightforward.

Third, there are many components that make up the muscular system and some show tremendous variation, both within a given animal (e.g. different fibre types) and between species. In particular, maximum velocity of shortening (V_{max}), rate of relaxation, rate of activation, mechanical gearing of the fibres, and myofilament lengths are known to show considerable variation. Some of these parameters vary in different fibre types, are temperature-dependent, change in response to physiological stress (e.g. thermal adaptation) and vary in different animals. Although much of the field of muscle physiology focuses on *how* (at the molecular level) this variation is achieved, this chapter is equally concerned with *why* this variation occurs in the first place. What are the rules (design goals) by which evolution sets particular values for these parameters in a given animal? Fourth, fish are the best animals in which to perform these studies because, unlike other vertebrates, their muscle fibre types are anatomically separated. This leads to many experimental advantages (see below).

Finally, a seemingly unique feature of muscle is that muscle function not only depends on what proteins are present, but also on the manner in which they are used. For instance, the force, power, and efficiency of muscle is not a fixed number, rather they depend on the V/V_{max} at which the muscle is shortening (illustrated in Figure 7). This last feature has two important consequences. First, unlike most biological systems, where the flux of a reaction can be increased by adding more enzyme, in muscle the right protein must be added to enable the animal to make movements effectively. For instance, increasing the amount of slow muscle will not enable the animal to move rapidly. Second, there are curves that define performance (the sarcomere length-tension curve, force-velocity curve, etc.) and there are optimal places on these curves for the muscle to work. Thus, whereas in most biological systems qualitative *a posteriori* reasons are usually given for the value of a given parameter, in the muscular system one can make an *a priori* hypothesis (i.e. the system is designed to conform to the constraints described by the curves), and then test it.

WHAT ARE THE RULES?

From cell physiology, we may anticipate there are two rules (Figure 1) that are followed when an animal muscular system is designed. During steady activation, the force that muscle generates depends on the amount of filament overlap between myosin and actin (or, more precisely, the number of myosin cross-bridges which can interact with actin sites; Gordon *et al.*, 1966). It would seem sensible for animals to vary the gear ratio (Δ body movement/Δ sarcomere length) of their muscle fibres and their myofilament lengths, so that no matter what movements the animal makes, the muscle would operate at optimal myofilament overlap (i.e. where the muscle generates near-maximal force). As such, gear ratio and myofilament lengths can be viewed as the *design parameters* (those components that can be varied during evolution). Optimal myofilament overlap can be viewed as a *design goal* or *design constraint* (i.e. the rule by which the variation in parameters is adjusted or what has to remain constant). As both design

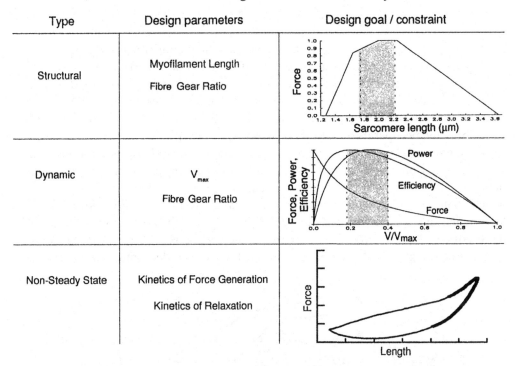

Type	Design parameters	Design goal / constraint
Structural	Myofilament Length Fibre Gear Ratio	
Dynamic	V_{max} Fibre Gear Ratio	
Non-Steady State	Kinetics of Force Generation Kinetics of Relaxation	

Figure 1. For three design considerations, the important design parameters (muscle properties which can be varied during evolution) and the *potential* design constraints are shown (system values that are kept constant). Empirical studies from my laboratory suggest that myofilament overlap and V/V_{max} are important design constraints; that is, the values of design parameters are set so that muscle operates only over the shaded portion of the curves where force, power and efficiency are maximal. Non-steady-state properties of muscle are also important and it appears that the rate of activation and relaxation are set to relatively slow values so as to provide maximum efficiency of power production.

parameters are anatomical features of the muscle, one at the organ level and the other at the molecular level, this can be viewed as a *structural* design consideration (Figure 1A).

There is also a dynamic design consideration which takes into account that muscle shortens during locomotion. The force that muscle generates is a function of V/V_{max}, where V is the velocity of shortening. In addition, and more importantly, the mechanical power that a muscle generates and the efficiency with which it generates the mechanical power are functions of V/V_{max} as well. Again we might anticipate that the muscular system would be designed in such a way that no matter what movement the animal makes, the muscle fibres operate over a range of V/V_{max} values (0·15-0·40) where the fibres generate maximal power at near maximum efficiency. Thus the design parameters V and fibre gear ratio are varied in such a way that they operate under the design constraint of V/V_{max} (Figure 1B).

It is important to emphasize that myofilament overlap and V/V_{max} are *potential* design goals, which are derived exclusively from experiments on *isolated* muscle. It was necessary to determine whether fish *actually use their muscles* over this narrow range of values during their full range of locomotion. Prior to the work of my laboratory, this had never been done.

DESIGN GOAL NO.1 - MYOFILAMENT OVERLAP

Because of the sliding filament structure of muscle, maximum force is generated over a fairly narrow range of sarcomere lengths (Gordon *et al.*, 1966). Researchers have generally assumed that the muscle is used only over those sarcomere lengths, but there has been relatively little evidence. If myofilament overlap is a design constraint, it raises an interesting question: how do animals produce such a wide range of movements if their fibres always operate at optimal myofilament overlap?

My laboratory's recent experiments on carp (*Cyprinus carpio* L.) provide the most extensive study of these issues. To swim, a fish must bend its backbone. By a combination of high-speed motion pictures, and anatomical and mathematical approaches which relate sarcomere length to backbone curvature, Rome *et al.* (1990) found that at low swimming speeds, the red muscle (Figure 2), which powers this movement, undergoes cyclical sarcomere length excursions between 1·89 to 2·25 μm centred around a sarcomere length of 2·07 μm (Figure 3). Further, it was determined from electron

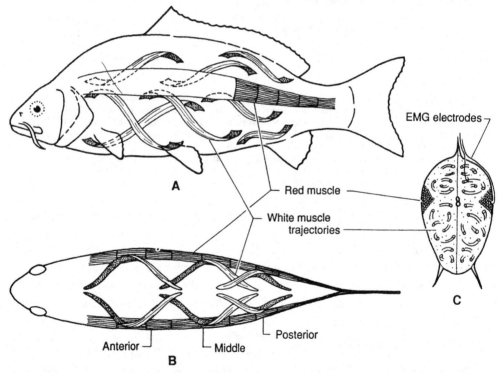

Figure 2. Longitudinal view (A), dorsal view (B) and cross-section (C) of carp. The red muscle represents a thin sheet of muscle just under the skin which extends to a depth of only 10% of the distance to the backbone (the cross-section of the red muscle is exaggerated for illustrative purposes). Because the red fibres run parallel to the body axis, sarcomere length excursion depends on both curvature of the spine and distance from the spine. The trajectories of the white muscle fibres shown in (A) and (B) are based on Alexander's (1969) description. The white fibres lie closer to the median plane than the red ones, and they run helically rather than parallel to the long axis of the body. Consequently, they shorten by only ~1/4 as much as the red ones for a given curvature change of the body (see text). Placement of electromyography (EMG) electrodes used to determine the activity of the red and white muscles are shown in (C). (Reproduced from Rome *et al.*, 1988.)

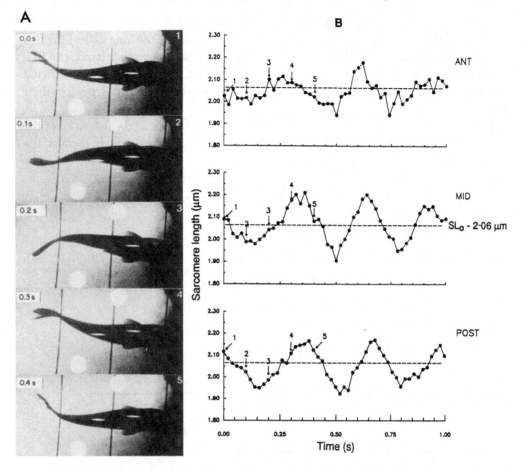

Figure 3. Typical record showing changes in sarcomere length in anterior, middle, and posterior positions of a fish swimming at 25 cm s⁻¹. Five frames (A) separated by 0·1 s are shown and numbers on the photographs correspond to data points in the 25 cm s⁻¹ graphs (B). Sarcomere length excursion (average difference between shortest and longest lengths measured in a sequence) was calculated from the amplitude of the graph. Muscle velocity was calculated from the slope of the graph. (Reproduced from Rome *et al.*, 1990.)

microscopy that thick and thin filament lengths of the red (1·52 μm and 0·96 μm) and white (1·56 μm and 0·99 μm) muscle in carp are similar to that in frog (Sosnicki *et al.*, 1991). Using the frog sarcomere length-tension relationship to approximate that of the red and white muscle, the red muscle was shown to be operating over a range of sarcomere lengths where no less than 96% maximal tension is generated (Figure 4; Rome & Sosnicki, 1991).

The most extreme movement carp make, the escape response (pictured in Figure 5) was then examined. This response involves a far greater curvature of the backbone than does steady swimming. *If* the red muscle were powering this movement, it would have to shorten to a sarcomere length of 1·4 μm, where low forces and even irreversible damage can occur (Figure 5B; Rome *et al.*, 1988; Rome & Sosnicki, 1991). Rather it is the

L.C. Rome

white muscle which performs the movement because the white muscle has a different orientation from the red. The red muscle fibres run parallel to the long axis of the fish (Figure 2) just beneath the skin. The white muscle fibres, on the other hand, run in a helical orientation with respect to the long axis of the fish. Alexander (1969), predicted that the helical orientation endowed the white fibres with two features crucial to their function.

First, the helical pattern enables the white fibres on one side of the fish to be at the same sarcomere length irrespective of the distance from the backbone. If, instead, the

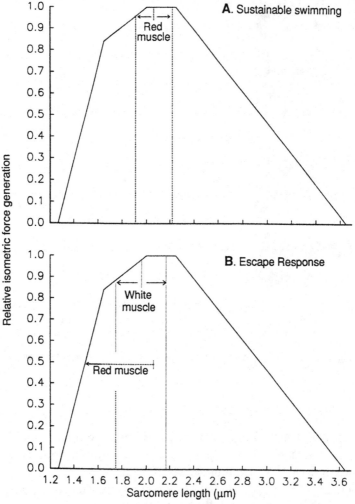

Figure 4. Design constraint 1. Myofilament overlap. During all movements, muscle is used at nearly optimal myofilament overlap. During steady swimming (A), the carp uses red muscle over a sarcomere length of 1·91-2·23 μm, where no less than 96% maximal tension is generated. If the red muscle had to power the more extreme escape response (B), it would have to shorten to 1·4 μm where it generates little tension and can be damaged. Instead the white muscle, which has a 4x greater gear ratio, is used. In the posterior region of the fish, the white muscle shortens to only 1·75 μm, where at least 85% maximal tension is generated. In the rest of the fish the excursion is smaller and the force higher. (Reproduced from Rome & Sosnicki, 1991.)

fibres ran parallel to the long axis, the ones far from the backbone would shorten to much smaller sarcomere lengths (and at a much faster V) than those close to the backbone. This would make it impossible for all the white fibres to work at optimal myofilament overlap and optimal V/V_{max}. Second, this helical orientation endows the

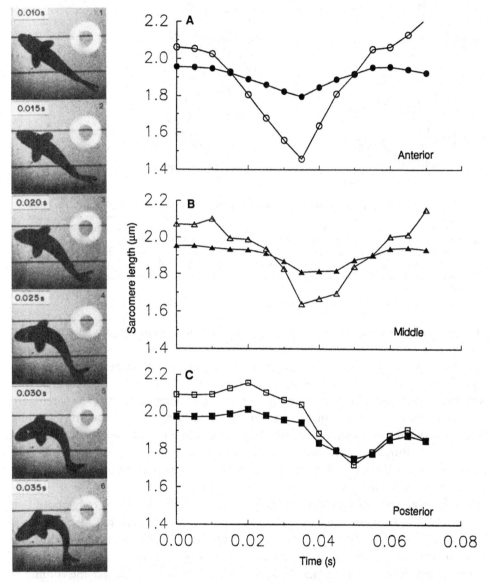

Figure 5. The startle response of a carp. A resting carp received a 100-ms, 150-Hz sound pulse through an underwater speaker in the aquarium about 30 cm from the fish. The response was filmed at 200 frames per second and six consecutive frames (separated by 5 ms) are numbered and shown on the left. The sarcomere length excursions of the white muscle in the anterior, middle and posterior positions are shown on the right (solid symbols). The sarcomere length excursion of the white muscle is greatest in the posterior, because here the backbone undergoes its largest curvature. Note that the open symbols show how far the red muscle would have to shorten *if* it were powering the movement. The red fibres *do not actually* shorten to the sarcomere length shown because they cannot shorten fast enough. (Adapted from Rome *et al.*, 1988 and Rome & Sosnicki, 1991.)

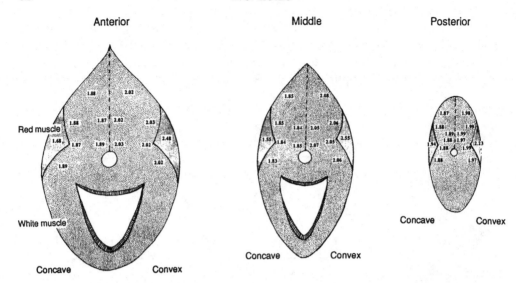

Figure 6. Sarcomere length distribution in steaks from the three positions along the carp. The cross sections of the anterior, middle, and posterior regions of a bent fish are shown as well as the sarcomere length at different points in the steak. Unlike the red muscle, the white muscle sarcomere length excursion did not depend on distance from the backbone. There was little variation in sarcomere length of the white muscle in a given steak and there was no consistent pattern to the variation, suggesting the white muscle operates as a unit. At all three positions, however, the sarcomere length of the red muscle is very different from that of the white. (Reproduced from Rome & Sosnicki, 1991.)

white muscle with a higher gear ratio, necessary to power extreme movements.

Rome & Sosnicki (1991) experimentally verified both predictions. Figure 6 shows cross-sections of bent fish, which illustrates two important points. First, the sarcomere length in the white muscle is the same both close to and far from the backbone, so that it works as a unit. Second, comparing the left and right sides of the fish, the sarcomere length difference (which is equivalent to the sarcomere length excursion) for the white muscle is much smaller than for the red. Hence the white muscle can produce a given backbone curvature for a much smaller sarcomere length excursion, and therefore has a higher gear ratio (N.B. in the context of fish swimming, gear ratio is most usefully defined as Δ backbone curvature/Δ sarcomere length).

Thus for a given backbone curvature, the white muscle undergoes only about one quarter of the sarcomere length excursion of the red (i.e. its gear ratio is four times higher). To power this most extreme movement of fish, in the worst case (posterior), the white muscle must shorten to a sarcomere length of 1·75 μm (Figure 4B). At this sarcomere length the muscle generates about 85% maximal force. Most of the volume of the white muscle (middle and anterior sections), however, does not shorten as much and thus generates even more force (95% for anterior and 92% for the middle). When the white muscle is used in less extreme movements (i.e. during fast swimming), the curvature of the backbone is not nearly as severe, and thus the white muscle generates nearly maximal force (Rome & Sosnicki, 1991).

As shown above, the myofilament overlap is never far from its optimal level, even in the most extreme movements (shaded portion of Figure 1A). It appears, therefore, that animals are designed in such a way that *no matter what the movement*, the muscles used generate nearly optimal forces. As such, myofilament overlap can be considered a design goal (i.e. a part of the system that is kept constant). Given the movements that fish need to make, evolution adjusts two design parameters (fibre orientation and myofilament lengths) such that the muscle fibres being used always operate at near-maximal myofilament overlap and force generation.

DESIGN GOAL NO. 2-V/V_{MAX}

Why V/V_{max} is important

V_{max} can vary greatly (over 1000-fold) and thus my laboratory has focused on this design parameter and on V/V_{max}. Different muscle fibres have different V_{max} values. Differences in V_{max} have been found in a given muscle at different temperatures, in different muscles in a given animal, in muscles from different animals, and, recently, with the advent of single muscle fibre techniques, in different muscle fibre types within the same muscle. To a first approximation, fibres with different V_{max} generate the same maximum isometric force per cross-section and have the same maximum efficiency, whereas the maximum power generated and rate of ATP splitting (N.B. one ATP is split for each cross-bridge going through the attachment-detachment cycle) in the fibre with a high V_{max} are considerably greater than in the fibre with a low V_{max} (Figure 7).

From Hill's (1938, 1964) work we know, however, that a muscle fibre's mechanical properties (force generation and power production) and energetic properties (ATP utilization and efficiency) are not simply a function of the fibre's V_{max}. They also depend on the V/V_{max} at which they are used. In the Huxley (1957) model, the mechanical and energetic properties of cross-bridges also depend upon V/V_{max} during shortening.

Figure 7 shows the force, power output, rate of energy utilization and efficiency of muscle with two different V_{max} values. For a given velocity of shortening, the force and mechanical power per cross-sectional area can be considerably higher in the fibre with a high V_{max} (at V_1, they are quite similar, whereas at V_2, they are quite different; Figure 7A,B). Therefore, to generate a certain force or power level, an animal need use fewer fibres with high V_{max} than fibres with low V_{max}. It would thus seem advantageous to have only muscle fibres with high V_{max}. There is, however, an energetic price paid for a high V_{max}. In Figure 7C the rate of ATP utilization in the fibre with a high V_{max} is considerably greater than in the fibre with a low V_{max}, at all velocities of shortening. There thus appears to be an adaptive balance between the mechanics and energetics of contractions. The balance can be best seen in Figure 7D. The fibres with low V_{max} are more efficient at low V (e.g. V_1). At higher velocities (e.g. V_2), however, the fibres with high V_{max} are more efficient. Thus to produce both slow and fast movements efficiently, the animal should use the fibres whose V_{max} is matched to the V at which it needs to shorten.

If V/V_{max} is in fact a design goal we would anticipate that a given muscle fibre type would be used over a range of V/V_{max} of about 0·15 to 0·40, where efficiency and power

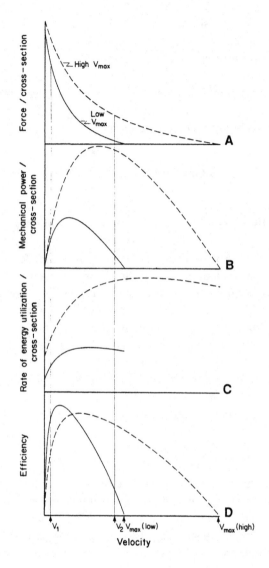

Figure 7. Relative force, power, rate of energy utilization and efficiency as a function of relative shortening velocity for a muscle with a high V_{max} (dashed curves) and a muscle with a low V_{max} (solid curves). Respective V_{max} values are shown on velocity axis. V_1 and V_2 are arbitrarily chosen examples of low and high shortening velocities. Values for curves are derived from heat, oxygen, and mechanics measurements on frog muscle (Hill, 1964; Hill, 1938; Rome & Kushmerick, 1983).

output are maximal. Thus the two design parameters, V_{max} and gear ratio, should be set so that during body movement the V falls within this range of V/V_{max}.

V/V_{max} had never been previously measured. To determine V/V_{max} one must determine (1) which muscle fibres are active during a given locomotory behaviour, (2) the V at which the fibres shorten, and (3) the V_{max} of the different fibre types. It is the first of these measurements which led my laboratory to work on fish. Because of the anatomi-

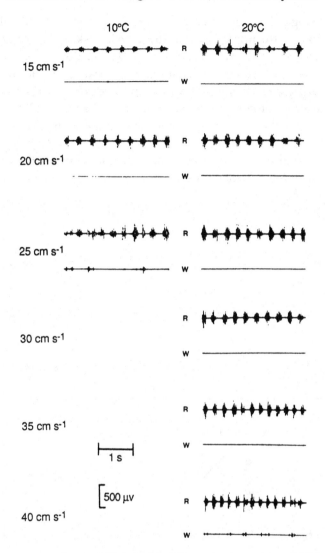

Figure 8. Electromyograms from red (R) and white (W) muscle of carp at 10 and 20°C. (Reproduced from Rome *et al.*, 1984.)

cal separation of their different muscle fibre types (Figure 2), it is possible to monitor activity of different fibre types by electromyography. For instance, we know from EMG at slow speeds that only the red fibres are active, and at fast speeds the white muscle fibres are recruited as well (Figure 8). Thus, by knowing the speed (and temperature) at which the fish is swimming, we know which muscle fibre types are actually being used.

This anatomical separation also helps in the other regards. First, it is relatively easy to isolate bundles of fibres which are *pure* in fibre type on which to measure V_{max} and other mechanical properties. This is a great aid because, in most other vertebrates, muscles are heterogeneous in fibre type and thus the only way to obtain a pure fibre type is using 'skinned' single fibres. Skinned-fibre experiments are not optimal for determining functional ability during locomotion because the skinning procedure makes it impossible to examine the kinetics of activation and relaxation. Second, as illustrated in regards to myofilament overlap, the anatomical separation enables us to define two separate gear ratios, one for the red muscle and one for the white, rather than the continuum of gear ratios found in other animals. Thus the value of V for each fibre type can be measured from the slope of the respective sarcomere length-time graph (Figures 3 & 5).

To test the importance of V/V_{max} as a design constraint, my laboratory has examined this parameter in three situations. Each situation was chosen to change either the

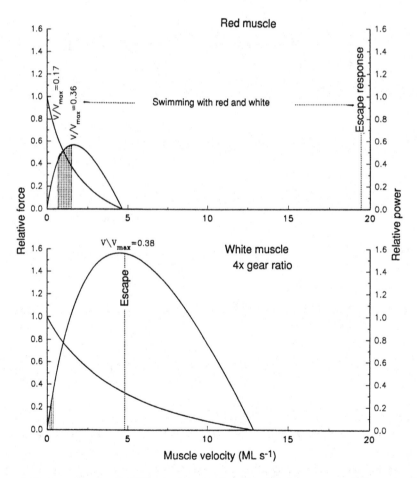

Figure 9. Design constraint 2. V/V_{max} in carp. During slow and fast movements the *active* fibres always shorten at a V/V_{max} of 0·17-0·38 where maximum power and efficiency are generated. During steady swimming (red muscle), the fibres are used at a V/V_{max} of 0·17-0·6 (A). The red fibres cannot power the escape response because they would have to shorten at 20 ML s^{-1} or 4x their V_{max}. The escape response is powered by the white muscle which need shorten at only 5 ML s^{-1} (V/V_{max} =0·38) because of its 4x higher gear ratio (B). The white muscle would not be well suited to power slow swimming movements, as it would have to shorten at a V/V_{max} of 0·01-0·03, where power and efficiency are low. Thus fast movements are obtained with fibres with a high V_{max} and a large gear ratio.

numerator, V, or the denominator, V_{max}, in order to see if there was a concomitant change in the other parameter to maintain a constant V/V_{max}.

Different fibre types in carp

The first question which was asked is why animals have different fibre types (i.e. different V_{max} within the same animal). Are the faster fibres used for faster movement (higher V) so that they operate at the same V/V_{max}?

As illustrated in Figure 9, Rome et al. (1988) found that the V_{max} of carp red muscle was 4·65 muscle lengths per second (ML s^{-1}) and the V_{max} of carp white muscle 2·5 times higher, 12·8 ML s^{-1}. During steady swimming the red muscle is used over ranges of velocities of about 0·7 to 1·5 ML s^{-1} (Figure 9A, shaded part of the curve, Rome et al., 1990). This corresponds to a V/V_{max} of 0·17-0·36 which is where maximum power is generated. At higher swimming speeds (higher V) the fish recruited their white muscle because the mechanical power output of the red muscle actually declines.

It is clear from Figure 9A that the red muscle cannot possibly power the escape response. To power the escape response, the red muscle would have to shorten at 20 ML s^{-1}, which it clearly can not do, as this is four times its V_{max}. But even if the white muscle were placed in the same orientation as the red (i.e. the same gear ratio), it could not either, because its V_{max} is only about 13 ML s^{-1}. However, because of its four-fold higher gear ratio, the white muscle need shorten at only 5 ML s^{-1} to power the escape response (Figure 9B), which corresponds to a V/V_{max} of about 0·38, which is where white muscle generates maximum power. The fact that all the white muscle (both close to the backbone and far from it) shortens by the same sarcomere length and thus with the same V and V/V_{max}, means that all the white muscle will generate maximum power. Thus the helical orientation of the fibres, which makes the white muscle work as a unit, is an important adaptation for maintaining an optimal V/V_{max}.

Thus the red and white muscle forms a two gear system which powers very different movements. The red muscle powers slow movements, while the white muscle powers very fast movements, both while working at the appropriate V/V_{max}. In terms of backbone curvature, the white muscle can produce 10-fold faster movements (2·5-fold higher V_{max} × 4-fold higher gear ratio). It is important to realize that the effectiveness of the white muscle depends on the product of its gear ratio and V_{max}. For instance, if the red muscle had the gear ratio of the white, it could not produce the escape response, nor could the white muscle, if it had the gear ratio of the red. What is needed is both the correct V_{max} and the correct gear ratio to produce very rapid movements.

As discussed above, the red and white muscles have a maximum V/V_{max} at which they can be used, but there also appears to be a minimum V/V_{max} at which they are used, which is also an important determinant of muscle function. For instance, if the white muscle does so well producing fast movements, why does the fish not have only one fibre type and let the white muscle power the slow swimming movements as well? The white muscle could certainly power slow swimming, but it is not used because its high V_{max} and four-fold higher gear ratio would make its V/V_{max} at slow swimming speeds, too low (i.e. 0·01-0·03, where the muscle is nearly isometric and efficiency is nearly zero; shaded portion of Figure 9B). At slightly faster swimming speeds, when the white muscle starts to be recruited to augment the power of the red, it still has the same problem. If the fish were to continue swimming steadily, the V of the white muscle would be only slightly faster than the shaded portion in Figure 9B, and the white muscle would still be shortening with too low a V/V_{max} to generate power efficiently. Under these circumstances, however, the fish employs 'burst and coast' swimming in which it makes rapid tail beats (i.e. short duty cycle) with the white muscle to obtain a high V and to keep an optimal V/V_{max}.

At very slow swimming speeds, (i.e. those below steady swimming, to the left of the shaded portion in Figure 9A), the carp also use the burst and coast pattern, but this time with the red muscle. This allows the fish's red muscle to make the normal sarcomere excursion in less time (shorter duty cycle) than the tail-beat period predicted for steady swimming. Thus the V (and V/V_{max}) will be higher, resulting in normal efficiencies. Thus carp appear to adopt a burst and coast pattern of swimming (red at very slow speeds, white at fast speeds), when the muscle fibre type being used would have too small a V/V_{max} to generate power with a high efficiency if the fish were swimming steadily.

In summary, given the constraints at both high V/V_{max} and at low V_l/V_{max}, to achieve a full repertoire of movement animals must have different fibre types with different V_{max}'s and different gear ratios. Thus V/V_{max} appears to be an important design goal or constraint (i.e. muscle operates over a narrow portion of the curve, 0·17-0·38, where maximum power is generated at high efficiencies; shaded portion in Figure 1B).

Different muscle temperatures in carp

Raising the temperature of a muscle increases its V_{max}. If V/V_{max} is an important design constraint, then the animal should use its muscle over a higher range of V at the higher temperatures. The V_{max} of carp red muscle is 1·6-fold greater at 20°C than at 10°C (Figure 10; Rome & Sosnicki, 1990). Rome et al. (1990, 1992a) found that at a given swimming speed, the V at which the muscle shortens is independent of temperature. Carp at 20°C, however, could swim to 45 cm s^{-1} with the red muscle exclusively, while at 10°C they could swim to only 30 cm s^{-1}. The corresponding V's were 1·6-fold higher at 20°C than at 10°C (2·04 vs 1·28 ML s^{-1} respectively), resulting in the same V/V_{max} (0·36) at the respective maximum speed at both temperatures (Figure 10, shaded portions). In addition, the fish at 10°C could swim steadily at lower speeds than at 20°C, corresponding to a 1·6-fold lower V (resulting in the same V/V_{max} at the lowest speed). Thus at both 10 and 20°C, carp use their red muscle over the same range of V/V_{max} (0·17-0·36). Conversely, V/V_{max} seems to set the operating range of the red muscle at different temperatures, and hence the swimming speed range. As such it is a powerful determinant of locomotory behaviour.

Fast swimming species (scup) vs slow swimming species (carp)

At a given temperature the marine scup (*Stenotomus chrysops* L.) can swim twice as fast with its red muscle as the fresh water carp (80 cm s^{-1} vs 45 cm s^{-1} at 20°C; Rome et al.,

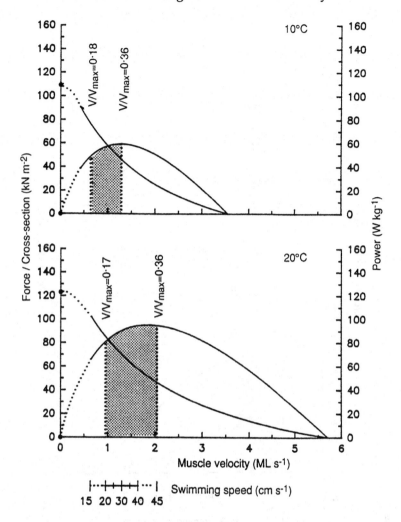

Figure 10. Influence of temperature on mechanical properties of carp red muscle and on their use during swimming. This figure shows the average force-velocity and power-velocity curves of carp red muscle at 10 and 20°C based on the results of Rome & Sosnicki (1990). As muscle shortening velocity during steady swimming was independent of temperature, the swimming speed axis has been placed on the graph as well. Thus during steady swimming, the curves provide the power, force and muscle shortening velocity as a function of swimming speed. The shaded regions represent the V during steady swimming with red muscle. The dotted vertical lines at each temperature, represent transition swimming speeds. At slower swimming speeds than that of the left-most line, the carp used 'burst and coast' swimming with red muscle. At higher swimming speeds than that of the right-most line, the white muscle is recruited and the carp used burst and coast swimming. For each temperature, the V/V_{max} at the transition points is given. Note that V/V_{max} over which the red muscle is used is the same at both temperatures. (Reproduced from Rome et al., 1990.)

1992a). It was anticipated that the scup's maximum V (i.e. the V while swimming at 80 cm s⁻¹) would be twice as high as the carp's maximum V (i.e. while swimming at 45 cm s⁻¹). If maintaining V/V_{max} is important, then the V_{max} of scup should be twice as high as the carp's as well.

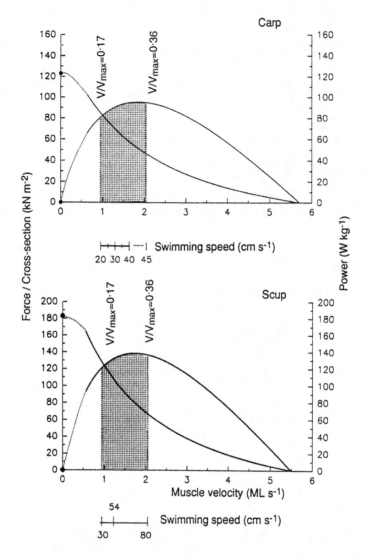

Figure 11. V/V_{max} of the fast swimming scup and slow swimming carp. The force-velocity and power-velocity curves are shown for the red muscle of both species at 20°C. Despite the fact that scup can swim to 80 cm s^{-1} with the red muscle while carp can only swim to about 45 cm s^{-1}, the V_{max} of both species are nearly the same. At the maximum swimming speed at which the red muscle is used (80 cm s^{-1} in scup and 45 cm s^{-1} in carp), it shortens at the same V (about 2·04 ML s^{-1}). Hence both species use their red muscle over the same range of V/V_{max}, but this occurs at higher swimming speeds (and higher tail-beat frequencies) in the scup. (Constructed with data from Rome & Sosnicki, 1990; Rome et al., 1990; Rome et al., 1992a,b.)

Remarkably, because scup employ a less undulatory style of swimming than carp (i.e. smaller backbone curvature and smaller sarcomere length excursion), at their respective maximum swimming speeds with red muscle, the V's at which the muscles shorten are equal (about 2·04 ML s^{-1}; Rome et al., 1992a). This finding led Rome et al. (1992b) to change their prediction. They now expected carp and scup should have the same V_{max}, which is exactly what they found. Hence, both the fast fish and the slow fish use their red muscle over the same V/V_{max} even though it corresponds to much higher swimming speed in scup (Figure 11). Two further points are of interest. First, fast-swimming fish do not necessarily need fast-contracting muscle. Second, scup should not necessarily be viewed as better-designed than carp because they can swim faster. Carp can swim steadily at much lower speeds than scup (Figure 11), which may be important to their life style (when a scup swims slowly it has to use its pectoral fins).

Summary of steady-state design constraints

Thus, in three different cases which might lead to changes in V/V_{max}, evolution has made adjustments to maintain V/V_{max} constant between 0·17 and 0·38, where power is maximal and where isolated muscle experiments suggest that near optimal efficiency is achieved. Although there have been no energetic experiments on carp and scup, those on dogfish muscle (Curtin & Woledge, 1991) show that, over the full range of V/V_{max} of 0·17 to 0·38, efficiency is at least 90% of its maximal value. These results show that V/V_{max} is an important design goal (Figure 1B). The way animals produce a wide range of movements is by using fibres with different V_{max} and with different gear ratios.

Thus animals use their muscles over a narrow range of myofilament overlaps and over a narrow range of V/V_{max} where muscle generates maximum force and maximum power with optimal efficiency (Figure 1A,B). Therefore during evolution three design parameters (gear ratio, V_{max} and myofilament lengths), appear to have been adjusted so as to obey these design goals, no matter what movement is made. Hence, these design goals appear to be two of the rules by which muscular systems have been put together.

NON-STEADY-STATE MECHANICAL PROPERTIES OF MUSCLE

Although V/V_{max} and myofilament overlap appear to be very important design goals, these muscle properties may not represent a complete description of muscle behaviour during locomotion. These design goals are based on the force-velocity and sarcomere length-tension curves, which are measures of *steady-state* mechanics of *maximally* activated cross-bridges, and hence do not take into account the fact that muscle is cyclically activated and relaxed during locomotion. Although these steady-state measurements are useful for studying the mechanisms of contraction, they may ignore important constraints on muscle performance during locomotion (Josephson, 1985a,b; Altringham & Johnston, 1990a,b; Marsh, 1990; Rome, 1990; Rome & Swank, 1992; Rome *et al.*, 1992b; Johnson & Johnston, 1991; Woledge, 1992).

Muscle generates positive work when it shortens while developing force, and produces negative work when it develops force during lengthening. The work output during a shortening-lengthening cycle which can be used to power swimming is the *net* work (positive minus negative work), which is graphically equivalent to the area contained within the force-length loop (Josephson, 1985a). Net power in cyclical contractions is described by the following formula: Net power output = (work done by $muscle_{shortening}$ - work done on $muscle_{lengthening}$) per cycle × cycle frequency. If muscle could instantaneously activate and instantaneously relax then the net work would be the product of the length change and force (from steady-state force-velocity curve) during shortening. However, because muscle does not instantaneously activate and relax, at cycling frequencies which fish use for swimming, there may be several important limitations on performance.

First, because the muscle cannot instantaneously relax (for our purposes, relaxation means that all the bridges are detached), the muscle may have to operate in such a way that *deactivation proceeds during shortening* so as to avoid high forces and negative work during subsequent relengthening. If so, the muscle can not remain completely activated during the shortening period. Hence, force production will be below that described by the steady-state force-velocity curve. Deactivation during shortening is caused by the nervous system cutting off the stimulus before the end of shortening, and by an intrinsic property of the muscle called 'shortening deactivation' which causes the force-generating ability of the fibre to be reduced during and following shortening (Edman, 1980).

Second, if the stimulus does not precede shortening by a sufficiently long time, or if the muscle is not stimulated for a sufficiently long duty cycle, then activation (which for our purposes includes attachment of the cross-bridges, force generation, and stretching of the series elastic component) may be incomplete during shortening. Hence, incomplete activation, or deactivating the fibre too early in the cycle, results in a lower positive work. In addition, incomplete relaxation results in increased negative work. From these considerations we can anticipate that (1) power output during swimming may involve a complex interplay between different muscle properties, and (2) the effect of activation and relaxation kinetics on the mechanical performance of the muscle depends greatly on the exact length change and stimulation pattern the muscle undergoes during swimming.

Thus, to assess the importance of non-steady-state properties of muscle and to understand how muscle is designed, my strategy has been to measure the exact length changes and stimulation pattern that occurs *in vivo* and then impose these on isolated muscle while measuring force and work production. Seven parameters completely define the length change and stimulation pattern: (1) oscillation frequency; (2) sarcomere length amplitude; (3) starting sarcomere length; (4) shape of sarcomere length change (i.e. triangular or sinusoidal wave); (5) phase of stimulation with respect to the length change; (6) stimulation duration; (7) stimulation rate. My laboratory has developed the techniques to measure six of these parameters and the seventh, stimulation rate, can be approximated by quantitative EMGs.

Rome *et al.* (1993) have recently applied this technique to swimming scup. Figure 12 shows the results of the overall experiment. They first swam fish at 80 cm s^{-1} and measured EMGs and muscle length changes of the red muscle at four places along the length of the fish (ANT1, ANT2, MID, POST). It was found that moving caudally along the length of the fish, the length change of the muscle becomes larger. Strains were ±1·5% of resting muscle length (resting sarcomere length was 2·10 μm) at ANT1 increasing to ±5·7% at the POST (Figure 12B). In addition, the EMGs had a longer duty cycle in the anterior of the fish than in the posterior (Figure 12A). This is due primarily to a large difference in the onset time of the EMG in the anterior compared to the

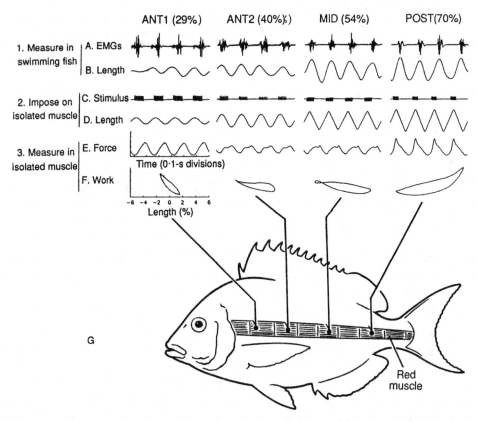

Figure 12. Length changes, stimulation pattern, force production and work output of red muscle of scup during swimming. Step 1 was to measure in a swimming (80 cm s⁻¹) fish the EMGs (A) and length changes (B) for the red muscle at four places along the length of the fish pictured in (G). Step 2 was to impose on muscle bundles isolated from these four positions the stimulation pattern (C) and length changes (D) that were observed during swimming. Step three was to measure in the isolated muscle the resulting force production (E) and work production (F). Note that the reason for the apparent discrepancy between traces A-B and C-D is that A-B represent records from one of the fish (tail-beat frequency = 6 Hz) while C-D represent the record for a muscle driven through the *average* swimming values (i.e. tail-beat frequency = 6·4 Hz). Traces A-E are all functions of time. Trace F is a plot of force produced against length changes where the area of the enclosed loop is the work produced during a tail-beat cycle. This value is much larger in the POST than in the ANT1 position. (Adapted from Rome *et al.*, 1993.)

posterior, whereas off-times were nearly simultaneous. Synchronization of EMGs with length changes showed that the EMGs precede the length changes by increasing amounts along the length of the fish towards the tail (Figures 12A,B & 13A,B).

To determine the mechanical performance of the muscle during this swimming behaviour, they drove red muscle bundles, isolated from the four positions on the fish, through the length changes and stimulation conditions they undergo during swimming and measured the resulting force and power the muscle generated. They found that despite the fact that the POST muscle is stimulated primarily during lengthening (Figure 13I), it generated large amounts of work. The anterior muscle, on the other hand, generated significantly less work under its *in vivo* conditions (Figures 12 & 13I,II).

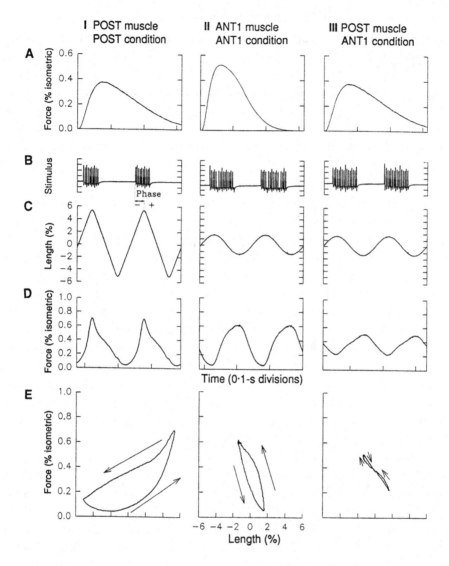

Figure 13. Mechanical properties of ANT1 and POST muscles. Columns I and II show a POST and an ANT1 muscle bundle undergoing the respective length changes and stimulation pattern that the muscle undergoes during swimming. By contrast column III shows a POST muscle undergoing the stimulation pattern and length changes that are encountered by the ANT1 muscle during swimming. Trace A shows the isometric twitch of the muscles in question. Traces B and C show the imposed stimulation pattern and length changes determined during swimming experiments. (Note the phase of the stimulus is defined with respect to maximum muscle length.) The resulting force is shown in trace D and the resulting work (area enclosed by loop) in E. A-D are functions of time, whereas E is force as a function of length. Note the large strain in the POST compared to the ANT1 and the much larger work produced in that muscle. Note also that relaxation is much faster in muscle undergoing shortening (D, caused by shortening deactivation) than in that being held isometrically (A). Finally note that relaxation is much faster in ANT1 muscle than POST (A). This permits the ANT1 muscle (IIE) to perform work under conditions where the POST muscle cannot (IIIE). (Adapted from Rome *et al.*, 1993.)

The cause of the low power generated in the anterior region of the fish is due not to an intrinsic deficiency of the anterior musculature, but rather to the specific length change and stimulation pattern that the anterior muscle undergoes. For instance, when the ANT1 muscle is driven through the length change and stimulation pattern that the POST muscle normally undergoes *in vivo*, it generates just as much power as the POST muscle. The main reason for the low power output in the anterior region is the small strain (*dl*). A small *dl* reduces work output (*Fxdl*) per tail beat. In addition, muscles undergoing small strains do not relax sufficiently fast. Relaxation is speeded up greatly by muscle shortening (shortening deactivation), but the shorter the strain, the smaller is this enhancement of relaxation (Josephson & Stokes, 1989; Altringham & Johnston, 1990; Rome & Swank, 1992).

The fact that the ANT1 muscle can produce any work at all during swimming represents an adaptation. For instance, when the POST muscle is driven through the length change and stimulation pattern the ANT1 muscle normally undergoes *in vivo*, it performs even worse than the ANT1 muscle and generates no net work (compare Figure 13II,III). The primary reason that the ANT1 muscle can generate significant power despite undergoing only a small strain (Figure 13II) is that it has the intrinsic ability to relax much more rapidly than the muscle in the posterior (Figure 13A). Because of its slow intrinsic relaxation rate, POST muscle going through the ANT1 conditions does not come close to relaxing between stimulus trains, whereas the ANT1 muscle relaxes almost completely (Figure 13IID,IIID). Although the mechanism for faster relaxation in the ANT1 muscle has yet to be explored, these results show that relaxation is malleable and that it can greatly effect power production by the muscle.

Although it had previously been proposed that the power for swimming is generated by the anterior musculature, and that the posterior musculature serves only to transmit force to the tail by lengthening contractions (Videler & Hess, 1984; van Leeuwen *et al.*, 1990), these results show that in fact the opposite is true. Most of the power generated during this type of swimming is generated by the *posterior* rather than the *anterior* musculature. This conclusion is further supported by additional kinematic analysis. During the period that the tail is sweeping to one side and generating mechanical power, the posterior muscle on that side of the fish is *shortening*, not *lengthening* (*lengthening* of the posterior muscle during this period would be necessary for the previously proposed hypothesis).

Generating power in the middle and posterior musculature seems to represent a superior design of the locomotory system because in this case all the power generated by the musculature can be used to power swimming. In addition, the posterior muscle appears to be optimized for power generation because, as discussed above, during these swimming movements it shortens over the portions of the sarcomere length-tension and force-velocity curves where it generates maximum power (Rome *et al.*, 1992a,b). In the case of the alternative hypothesis, where the posterior muscle would lengthen and perform negative work, then a portion of the positive power generated by the anterior musculature would presumably be lost as heat in the posterior. Avoiding this loss may be essential because the red muscle makes up less than 4% of the mass of the fish, but powers all of steady swimming.

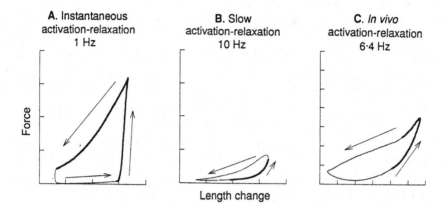

Figure 14. Work loops performed by scup red muscle during oscillatory contractions at 1, 10 Hz and *in vivo* conditions. The heavy lines on the work-loops correspond to the stimulus being on. Note that at 1 Hz (A), the stimulus is turned on just prior to shortening and is turned off just prior to the end of shortening. Hence activation and relaxation can be viewed as instantaneous and have little effect on muscle performance. At 10 Hz (B), however, the stimulus is started well before shortening begins and it also ends before shortening even begins. Thus *if* scup swam at this frequency, the kinetics of activation and relaxation would impinge significantly on muscle performance, and hence we would conclude that evolution had set the activation and relaxation rates to be slow. During actual locomotion (swimming at 80 cm s^{-1} with a tail-beat frequency of 6·4 Hz), the activation and relaxation processes have a large influence on power production and thus we conclude that evolution sets these processes to be relatively slow.

WHAT IS THE THIRD DESIGN GOAL?

Although this approach of reproducing in isolated muscle the length change and stimulation pattern which the muscle undergoes during locomotion has given us new insights into how fish swim, its real power is to enable us to develop and test how evolution has set the rate of activation and relaxation in a given muscle system.

There are two extreme cases by which evolution could have set the rates of activation and relaxation. Figure 14A,B shows the work loops from scup red muscle going through a ±5% length change at 1 Hz and 10 Hz while the phase and stimulus duration have been set for optimal power generation. If the scup never swam with a tail-beat frequency greater than 1 Hz, then we would conclude that the muscle is designed to provide instantaneous (i.e. the process is completed over a small portion of the cycle) activation and instantaneous relaxation and that activation-relaxation processes had little effect on the mechanical properties. The reason is that the muscle is stimulated (represented by the thickened line) for 50% of the cycle, including the entire shortening phase. Thus it should be maximally activated during shortening and generate maximum positive work. In addition, the muscle appears to relax instantaneously at the end of shortening, and no negative work is done during relengthening.

If on the other hand, scup swam with a 10 Hz tail-beat frequency, we would conclude that muscle activation-relaxation processes severely limit power production, and evolution sets the activation-relaxation rate to be very slow (i.e. the process takes a large fraction of the cycle). In this case the muscle is only stimulated for 25% of the duty cycle. To ensure that the muscle has time to relax prior to relengthening, the muscle must be stimulated well in advance of shortening, and the stimulation ends even before the shortening starts. Thus, although generating some force, the muscle is actually in the process of relaxing throughout shortening, and this reduces positive work output. In addition there is substantial negative work during lengthening which further reduces the net work production.

Thus, the crucial question is, at the frequencies scup actually use, did evolution set the activation-relaxation rate to be instantaneous or to be slow? Scup swimming at 80 cm s^{-1} use a tail-beat frequency of 6·4 Hz. The work loops under these conditions (Figure 14C) show that the muscle is designed with a relatively slow activation-relaxation rate. Thus the muscle is stimulated for only 27% of the duty cycle and is relaxing during most of the shortening phase. Thus during locomotion, activation-relaxation has a large effect on work production.

Intuitively one might have thought that it would be better for the muscle to have evolved instantaneous activation-relaxation at the frequency the scup use their muscles because it would generate more power. Thus we might expect the work loop at 6·4 Hz to resemble the one in Figure 14A. However, to enable the muscle to relax 'instantaneously' at 6·4 Hz, the Ca^{2+} pumping rate would have to be greatly increased, which would increase the energetic cost of the contraction. Thus it is likely that evolution sets the activation-relaxation rate to be slow so that the mechanical power can be generated most efficiently. This hypothesis can be verified empirically by measuring efficiency of power production while the muscle is undergoing its *in vivo* length change and stimulation pattern.

CONCLUSION

This chapter has shown that although there is tremendous variation in the components of muscular systems, these components are put together in a given muscular system following three relatively simple rules. First, myofilament lengths and fibre gear ratios are set so that no matter how extreme the movement, muscles operate over optimal myofilament overlap. Second, V_{max} and fibre gear ratio are set so that no matter how fast the movements are made, the active fibres always operate at optimal V/V_{max}. Finally, the rate of activation and relaxation are set to be relatively slow, which may permit the muscle to generate power most efficiently.

This work was supported by grants from the National Institutes of Health and National Science Foundation

Chapter 7

How do fish use their myotomal muscle to swim?
In vitro simulations of *in vivo* activity patterns

JOHN D. ALTRINGHAM

Department of Pure and Applied Biology, The University of
Leeds, Leeds, West Yorkshire, LS2 9JT, UK

In a number of laboratories, the mechanical properties of myotomal muscle are being investigated under conditions which approximate to those believed to be operating *in vivo* during steady swimming. The aim of these studies is to learn how myotomal muscle is used to generate swimming movements. To perform these experiments, the muscle length changes during steady swimming (the strain cycle) and the pattern of muscle activation (activation cycle) must be known. Electromyographical (EMG) recordings from several points along the length of swimming fish have provided information on the activation cycles. Strain cycles of superficial muscle fibres have to date been determined primarily by calculation from the changes in shape of the fish plan outline. Simultaneous EMG and kinematic analyses yield the relation between strain and activation cycles. By imposing cyclical strains on isolated muscle, and stimulating at different phases in the strain cycle, a wide range of possible *in vivo* situations can be simulated, and muscle performance under these conditions can be compared to values under conditions which yield, for example, maximum power, work or stress. Valuable insights are thus gained into design constraints and the mechanisms by which muscle function has been optimized by selective pressures.

INTRODUCTION

An answer to the question posed by this paper is still some way in the future. However, recent investigations which draw on both *in vivo* studies of swimming fish and *in vitro* experiments on isolated muscle fibres, are shedding some light on the problem. To understand the functional role of myotomal muscle, we need to know the conditions under which it is operating *in vivo*, and its mechanical performance under

Maddock, L., Bone, Q. & Rayner, J.M.V. (ed.). *Mechanics and Physiology of Animal Swimming.*
© 1994. Cambridge University Press.

these conditions. The conditions can be determined only from studies of swimming fish, but measuring muscle performance in a live fish is not yet possible. One obvious approach is to impose the *in vivo* conditions on isolated muscle fibres, study their responses, and attempt to fit the results into some sort of model. First, we need to know what is happening *in vivo*. Most fish swim primarily by the propagation of lateral oscillations down the body, and this paper is concerned only with this form of swimming. The nature of these oscillations changes among species, and each species is loosely placed in one of three classes, anguilliform, carangiform or tunniform. At one extreme, anguilliform fish have a little more than one large-amplitude wave on the body at any one moment, and generate thrust along most of their length. Carangiform swimmers generate most of their thrust at the tail-blade, with a single wave on the body. In stiff-bodied tunniform fish, only the caudal region oscillates. The most extensively studied group of fish are carangiform swimmers, and I will concentrate on these.

Steady swimming speed increases with increased tail-beat frequency and with the recruitment of more fibres within the lateral (myotomal) muscle (Bone *et al.*, 1978; Johnston & Moon, 1980; Rome *et al.*, 1990), but patterns of body movement (and hence muscle strain) and muscle activation remain essentially constant (Hess & Videler, 1984; Videler & Hess, 1984; van Leeuwen *et al.*, 1990; Wardle & Videler, 1993). Recent kinematic and electromyographical (EMG) studies (Hess & Videler, 1984; Williams *et al.*, 1989; van Leeuwen *et al.*, 1990; Wardle & Videler, 1993) provide a description of muscle strain and activation cycles during swimming and are a useful working basis for studies on isolated muscle fibres. Put simply, during steady swimming myotomal muscle undergoes cyclical length changes, and during a particular part of each strain cycle, related to its functional role, the muscle is activated. This type of activity lends itself to analysis using the work-loop technique. Originally used on asynchronous (myogenic) insect flight muscle, the technique was modified by Josephson (1985a) for use on synchronous (neurogenic) insect flight muscle, and its potential for the study of all types of synchronous muscle was recognized. By manipulating the frequency, amplitude and waveform of the strain cycle, and the number and timing of muscle stimuli in relation to the strain cycle, a wide range of possible *in vivo* conditions can be mimicked. This paper reviews the contribution made by recent work-loop experiments to our understanding of fish muscle function: other approaches are discussed by other contributors to this volume (see Rome, chapter 6; Wardle & Videler, chapter 8).

FAST (WHITE) AND SLOW (RED) MUSCLE FIBRES

As tail-beat frequency and swimming speed increase, there is a sequential recruitment of myotomal muscle from superficial to deep muscle fibres (Johnston *et al.*, 1977; Bone *et al.*, 1978; Johnston & Moon, 1980; Rome *et al.*, 1990). At slow, sustainable swimming speeds, only the slow muscle fibres close to the lateral line are active. The cost of locomotion is low at these speeds, and this fibre type typically makes up just a few percent of the total myotomal muscle mass. The power required for swimming increases rapidly with increasing speed ($\propto L^{2.5}$, Webb, 1978c), and requires the recruitment of fast muscle fibres, which form the bulk of the myotomal muscle. Swimming speed is largely determined by tail-beat frequency, since the distance moved during one beat is an essentially constant proportion of body length (Webb *et al.*, 1984). In elasmobranchs, and some primitive teleosts, it appears that fast fibres are not used during sustained/intermediate speed swimming, but only during short bursts of rapid activity (Bone *et al.*, 1978). In many teleosts, evidence suggests that the more superficial fast fibres do have a role to play in sustained/intermediate speed swimming (Johnston *et al*, 1977; Bone *et al.*, 1978; Johnston & Moon, 1980), although this may not be the case in all species (*e.g.* Rome *et al.*, 1985). How do the mechanical properties of the fast and slow muscles measure up to these requirements? Only one work-loop study has compared fast and slow muscle fibres from the same species. In the sculpin, *Myoxocephalus scorpius* L. (Altringham & Johnston, 1990a) fast muscle produced 30 W kg^{-1} at 6 Hz, slow muscle only 8 W kg^{-1} at 2 Hz, under conditions yielding maximum power output (Figure 1). These results make qualitative sense, in that the slow fibres yield their maximum power output at a frequency of 2 Hz, in the middle of the steady swimming range of sculpin (1-3·5 Hz), and below that at which the fast fibres are first recruited (4 Hz, Johnston *et al.*, 1993). The fast fibres develop higher power at higher frequencies, consistent with their use at higher tail-beat frequencies (4-9 Hz, Johnston *et al.*, 1993). A study of maximum power-generating ability is the most appropriate in this case, as the fast muscle fibres, at least, appear to be operating under conditions which yield large net positive work in each tail beat (Johnston *et al.*, 1993).

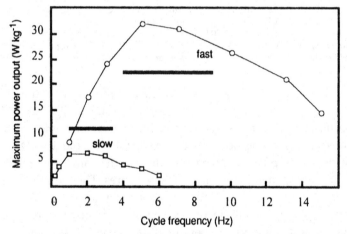

Figure 1. Maximum power output *vs* cycle frequency curves from representative preparations of fast and slow myotomal fibres from the sculpin, *Myoxocephalus scorpius*, at 3·5°C (based on Altringham & Johnston, 1990a). The horizontal bars show the range of frequencies over which the two fibre types are recruited during swimming (data from Johnston *et al.*, 1993).

Tang & Wardle (1992) calculated the power requirements of salmon swimming at their maximum sustainable swimming speed (at 12-16°C), using hydrodynamic models. Assuming that only the slow fibres were active at this speed, they estimated that the figure of 8 W kg^{-1} obtained for sculpin, with the available slow muscle mass, was sufficient to power sustainable swimming. The value for sculpin slow muscle was based on fibre volume, whereas the estimate of muscle mass made by Tang & Wardle

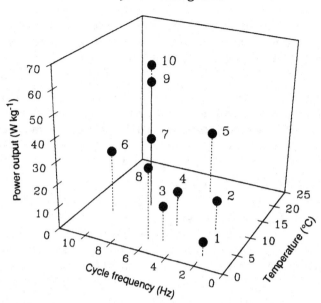

Figure 2. 3-D graph of maximum work-loop power in relation to temperature and cycle frequency, two of the major determinants of power output. Power increases with both temperature and frequency. 1, slow muscle, sculpin, *Myoxocephalus scorpius* (Altringham & Johnston, 1990a); 2, slow muscle, scup, *Stenotomus chrysops* (Rome & Swank, 1992); 3, fast muscle, cod (35 cm), *Gadus morhua* (Moon *et al.*, 1991); 4, fast muscle, cod (60 cm, Anderson & Johnston, 1992); 5, slow muscle, scup (Rome & Swank, 1992); 6, fast muscle, cod (10 cm, Anderson & Johnston, 1992); 7, fast muscle, saithe, *Pollachius virens* (0·65 BL, Altringham *et al.*, 1993); 8, fast muscle, sculpin (Altringham & Johnston, 1990a); 9, fast muscle, saithe (0·50 BL, Altringham *et al.*, 1993); 10, fast muscle, saithe (0·35 BL, Altringham *et al.*, 1993). Values for sculpin fast muscle similar to those reported by Altringham & Johnston (1990a) have been found by Johnson & Johnston (1991) and Johnston *et al.* (1993).

(1992) was derived from muscle volume, which includes non-contractile elements. Although the mass-specific muscle power used by Tang & Wardle (1992) should therefore be reduced by about 20-40% to account for the non-contractile component (unpublished observations), the slow muscle would still be able to power sustainable swimming. The power output of sculpin slow muscle will in any case probably prove to be at the lower end of the measured range for slow fibres. The sculpin is a benthic fish with a low tail-beat frequency range, and the experiments were made at 3°C. Power output from the work-loop technique, of a wide range of muscle fibre types increases with both frequency and operating temperature (Stevenson & Josephson, 1990; James *et al.*, 1994). Values obtained from fish muscle are shown in Figure 2. Muscles which can be driven to higher cycle frequencies tend to have higher power outputs, and come from fish with higher tail-beat frequency ranges (Altringham & Johnston, 1990b; see below), and high temperature facilitates operation at higher frequencies by increasing the rate of many contractile processes.

Slow fibres have been studied in only one other species, the scup, *Stenotomus chrysops* L., a fish of similar size to the sculpin, but with a more streamlined body and higher tail-beat frequency range (Rome & Swank, 1992; Rome *et al.*, 1992a). Maximum power was 13 W kg[-1] at 10°C (2·5 Hz) and 30 W kg[-1] at 20°C (5 Hz), comparable to the sculpin with allowance for the higher temperatures and slightly higher optimal frequencies. Rome &

Swank (1992) noted that at 20°C, under simulated *in vivo* conditions of strain and activation, isolated muscle generated close to maximum power output. However, at 10°C, maximum power of isolated muscle was obtained at a frequency lower than that used by the fish. The reason for this is unknown, but one suggestion is that an acclimation response may be responsible for the discrepancy.

Estimates of maximum fast-muscle-fibre power output using work loops are more numerous: all fall in the range 15-65 W kg^{-1}, with power tending to increase at higher temperatures and cycle frequencies (Figure 2). As yet, there has been no attempt to relate these to the power requirements of swimming. If it is assumed that all the fast muscle is active at maximum swimming speed, then the comparison should be possible; at lower speeds an unknown proportion of the fast muscle will be active.

SCALING EFFECTS

As fish size increases within a species, or between species with similar swimming modes, the tail-beat frequency range used and tail-beat amplitude decrease (Bainbridge, 1958; Archer & Johnston, 1989). Fish around 0·1 m in length are capable of swimming at up to 25 body lengths s^{-1}, but 1-m long fish rarely reach 4 body lengths s^{-1} (Wardle, 1975), due to their lower tail-beat frequencies. If evolution has resulted in optimization of muscle properties for locomotory function, then scale-dependent changes in locomotory patterns should be reflected in the properties of the myotomal muscle. Wardle (1975) found that the twitch contraction time of myotomal muscle blocks increased with increasing body size in a range of North Sea teleosts. Maximum swimming speed could be predicted from twitch contraction time and stride length, suggesting that the former is an important functional design constraint. Archer *et al.* (1990) also found a similar, positive relationship between body size and twitch contraction time in a nerve-muscle preparation from the cod, *Gadus morhua* L. In contrast, the velocity for maximum power output of fast fibres from the dogfish, *Scyliorhinus canicula* L., was independent of body size (Curtin & Woledge, 1988). This result is at first surprising, but contraction velocity is only one of many parameters which determine muscle power output (Altringham & Johnston, 1990b; Altringham *et al.*, 1993). The absence of a strong length dependence in any one mechanical parameter might therefore be expected.

In a swimming fish, muscle performance is determined by the complex interaction of many parameters, principally activation and relaxation, force-velocity and length-force relationships, shortening deactivation and force enhancement by stretch. Any change in muscle performance seen with increasing body size could be brought about by changes in one or more of these properties. A logical first step towards an understanding of how myotomal muscle has evolved to function in different-sized fish is to study its performance under simulated *in vivo* conditions. In cod, the frequency yielding maximum power decreased with increasing body size (Figure 3) ($\propto L^{-0.52}$, Altringham & Johnston, 1990b; $\propto L^{-0.59}$, Anderson & Johnston, 1992; in the range 10-70 cm body length, L). Mass-specific power output decreased from 28 to 16 W kg^{-1} ($\propto L^{-0.29}$), but stress was independent of body length (Anderson & Johnston, 1992). At least part of the shift to lower optimum cycle frequency can be explained by the longer twitch contraction time

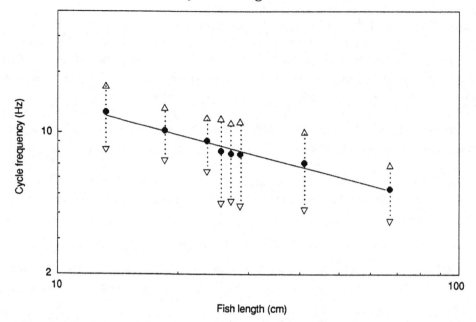

Figure 3. A log-log plot of frequency for maximum oscillatory work *vs* fish length for cod fast muscle fibres at 4°C. Bars indicate the range of frequencies over which the muscle generates more than 90% of peak power. The mid-points can be described by the equation: $f_{opt} = 46 \cdot 8 \, L^{-0 \cdot 52}$, where f_{opt} = frequency for maximum power output, and L = body length.

(Wardle, 1975; Archer *et al.*, 1990), which in the cod is due primarily to a slower relaxation (Altringham & Johnston, 1990b).

There are two obvious directions in which this work should proceed. The first looks outwards, to the performance of the whole animal. Isolated muscle properties must be related to the performance of the muscle in the fish. Kinematic data for the cod (Videler & Wardle, 1978) have not been used in this context yet. In addition, the hydrodynamic power requirements of the whole fish should be related to the power output of the muscle. The second looks inwards, to the muscle itself. The work-loop technique treats the muscle as a black box, and overall performance is measured, without a full understanding of how the machine works. Activation and relaxation, force-velocity and length-force relationships, shortening de-activation and force enhancement by stretch must all be quantified, and the results used to model *in vivo* function. Preliminary comparisons between work loops generated by such models, and those derived experimentally, have been very encouraging (van Leeuwen *et al.*, 1991).

THE ROLE OF FAST MUSCLE FIBRES IN DIFFERENT LOCATIONS ALONG THE BODY

The emphasis so far has been on muscle operating under conditions of maximum power output. It has become clear from kinematic/EMG investigations that during steady swimming at least, much of the myotomal muscle is operating under rather different conditions.

Electromyographical and kinematic data are particularly extensive for the saithe (*Pollachius virens* L.), a fast pelagic swimmer (Hess & Videler, 1984; Wardle & Videler, 1993). In steady swimming, waves of body curvature travel down the fish. The strain (length change) cycle of superficial lateral muscle is essentially sinusoidal. Waves of muscle activation, alternating from left to right sides, travel faster than those of body curvature, leading to systematic phase differences between strain and activation cycles along the body, suggesting a change in muscle function. Electromyographical activity in rostral myotomes is recorded primarily during muscle shortening, and in caudal myotomes during both lengthening and shortening. Similar results were obtained for superficial slow fibres of the carp by van Leeuwen *et al.* (1990). The results from carp were used to model mathematically the behaviour of slow myotomal muscle at different locations in the trunk, using parameters from other vertebrate muscle. Van Leeuwen *et al.* (1990) predicted that during steady swimming the myotomal muscle would have both positive and negative work components. Anteriorly the positive component would dominate and power output would be near maximal, but the negative component would become increasingly large towards the tail, with net work close to zero in the anal region, and negative near the tail (van Leeuwen *et al.*, 1990).

Another finding which appeared to be relevant was that the twitch contraction time of fast myotomal muscle blocks increased towards the tail in several teleost species (Wardle *et al.*, 1989). No hypotheses had been put forward to explain this observation, nor had it been confirmed on isolated fibres.

Altringham *et al.* (1993) studied the mechanical properties of superficial fast muscle fibres from three locations in saithe (*Pollachius virens*), 0·35, 0·5 and 0·65 BL (body lengths) from the head, and close to the lateral line. Muscle became slower towards the tail: time to peak twitch force and overall twitch contraction time increased, and tetanic fusion frequency decreased, but stress was constant. Using the work-loop technique, under conditions yielding maximum power output, the power *vs* frequency relation of saithe fast muscle shifted to lower frequencies towards the tail. This resulted in significantly higher power outputs for rostral relative to caudal muscle at high frequencies (12-18 Hz) (Figure 4). Maximum mass-specific power output of rostral muscle was twice that of caudal muscle (63 and 31 W kg^{-1} respectively).

Using the kinematic and EMG data of Hess & Videler (1984) and Wardle & Videler (1993), Altringham *et al.* (1993) performed experiments in which conditions approximating those *in vivo* were imposed on fibres from the three locations. Strain amplitude of superficial fibres in saithe at 0·35 BL (rostral) was ±3% l_o during steady swimming (Hess & Videler, 1984). Maximum power output at all cycle frequencies was obtained when the first stimulus was given at a phase shift of 30-40°, and the duration of the stimulus burst was 100-130° (one cycle = 360°, 0° = muscle lengthening through l_o, length for maximum tetanic force). This was the same phase shift between the strain cycle and the onset of EMG activity and the same EMG duration observed at all tail-beat frequencies in a swimming fish (Wardle & Videler, 1993). Under these conditions almost all force generation occurred during the shortening phase of the cycle, and the muscle therefore performed largely positive work. Caudal myotomes (0·65 BL) operate under very different conditions from rostral myotomes *in vivo*. Muscle strain amplitude

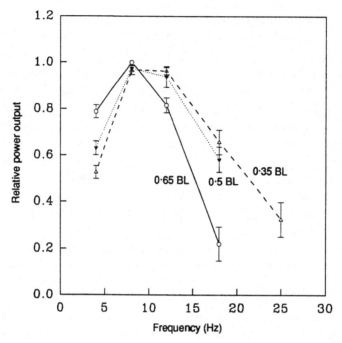

Figure 4. Relative maximum power output (mean ±SE) of saithe superficial fast muscle fibres plotted against cycle frequency at 12°C. Data from myotomes 0·35, 0·5 and 0·65 body lengths (BL) from the rostral tip, normalized to maximum at 8 Hz. (From Altringham et al., 1993.)

was ±6% l_o (Hess & Videler, 1984), the phase shift for the onset of EMG activity was 350°, and EMG burst duration 100° (Wardle & Videler, 1993). Muscle is activated when shorter than l_o, and in the *in vitro* simulations all the rising phase of force generation occurred while the muscle was being stretched. Muscle stiffness/force was maximal under these conditions. For mid-point myotomes, power was maximal at a phase shift of 0-30°: as with rostral myotomes a phase range similar to that predicted to be used *in vivo*, suggesting that under *in vivo* conditions muscle at 0·5 BL generates largely positive work during the tail beat.

Figure 5A shows force and strain records for muscle from 0·35, 0·5 and 0·65 BL, all operating under estimated *in vivo* conditions for steady swimming. Strain cycles of the 0·5 BL and 0·65 BL preparations lag that of 0·35 BL by 55° and 110° respectively, as *in vivo* (Wardle & Videler, 1993), due to the rostral to caudal propagation of the bending wave. Peak power output of muscle from 0·35 BL coincides with peak force/stiffness of muscle from 0·5 and 0·65 BL. In Figure 5B instantaneous power output is plotted for all three preparations over the middle two cycles. It can be seen that there is a sequential recruitment of myotomes as power generators during a tail beat, with the more caudally placed myotomes initially acting to transmit the power towards the tail. Positive power output is preceded by a period of negative work, during which time the muscle is at its maximum stiffness to resist stretch. This observation is consistent with the finding that most of the thrust in carangiform swimming comes from the blade of the caudal fin (Lighthill, 1971; Hess & Videler, 1984). Some of the energy stored by the stretching of

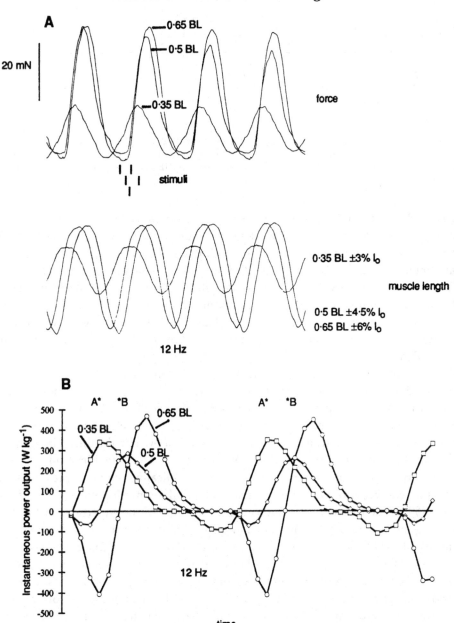

Figure 5. (A) Strain and force records of the middle 4 of 8 oscillatory cycles, for saithe fast muscle preparations from 0·35, 0·5 and 0·65 BL from the rostral tip. Muscle operating under predicted *in vivo* conditions at a cycle frequency of 12 Hz. Stimulus phase-shift degrees, number of stimuli and strain percentage l_o) were: 30, 2 and 3 (0·35 BL); 15, 2 and 4·5 (0·50 BL); 330, 1 and 6 (0·65 BL). Strain cycles of 0·5 BL and 0·65 BL preparations lag that of 0·35 BL by 55° and 110° respectively, as *in vivo*. Stimulus timing is indicated below the force record for the second cycle. Fish length, 29 cm. (B) Instantaneous power output over the middle two cycles shown in (A). A*, time of peak forces at 0·35 BL. *B, time of peak force at 0·5/0·65 BL. Force at 0·5 usually peaks slightly earlier than that at 0·65 BL. Muscle from 0·35 and 0·50 BL performs largely positive work. Positive work at 0·65 BL is preceded by a substantial period of negative work production, and mean power output is close to zero. (From Altringham *et al.*, 1993).

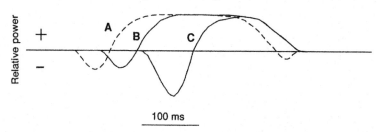

Figure 6. Calculated relative power output during a tail beat of steady swimming for slow myotomal muscle of the carp at positions approximating to those studied in the saithe by Altringham *et al.*, 1993). A, B and C ≈0·35, 0·5 and 0·65 BL from the rostral tip. (Based on van Leeuwen *et al.*, 1990, with permission).

active muscle is returned, since the force is higher throughout shortening than it would have been without prestretch. Since work is force multiplied by distance moved, muscle performs more positive work after a prestretch than without it. At 0·65 BL this force enhancement by stretch (Cavagna *et al.*, 1985) may be necessary since the smaller cross-sectional area of muscle at this location (about 50% of that at 0·35 BL, Videler & Hess, 1984) cannot generate the necessary stress by any other means. However, we do not know what role connective tissue may play in this region. These results show a striking similarity to the computer-simulated data published by van Leeuwen *et al.* (1990), and shown in Figure 6.

Figure 7 shows a schematic representation of the events of a single tail beat for the saithe, based on Altringham *et al.* (1993). The figure shows progressive fish outlines and midlines from above through one complete tail-beat cycle. The phase value on the left refers to the phase of the lateral curvature at the tail tip, *i.e.* muscle strain at the tip, if the tip had some muscle. Electromyographical activity is shown by the thick black line on either side of the midline (from Wardle & Videler, 1993). The dotted line indicates that EMG activity stops at that moment. Electromyographical activity occupies the same phase and relative duration within a tail beat at all swimming speeds (Wardle & Videler, 1993). The hatched areas indicate that segment of the superficial fast muscle generating more than 50% of the maximum force developed in the cycle, for a fish swimming at 18 Hz. The more rostral hatched areas indicate shortening muscle operating in the 'power phase': generating a large, positive mean power output. Caudally placed hatched areas represent muscle operating in its 'transmission phase': that is, active during lengthening, generating large forces, and transmitting the power generated by more rostral muscle towards the tail. Partial hatching is shown where the muscle force is falling below 50% of maximum. The timings of muscle stimuli, peak forces, and the transition from shortening to lengthening from the isolated muscle experiments are given. The predicted time of maximum tail-blade force and maximum bending moment (Lighthill, 1971; Hess & Videler, 1984; Tang & Wardle, 1992) is shown as a broad zone around 360°/0°.

Early in the right-ward tail sweep (tail-tip phase = 180°), power is generated by the most rostral right-side myotomes, and this power is transmitted towards the tail-blade by stiffened myotomes placed more caudally. Because of the phase differences between the caudally travelling waves of muscle activation and bending (Hess & Videler, 1984;

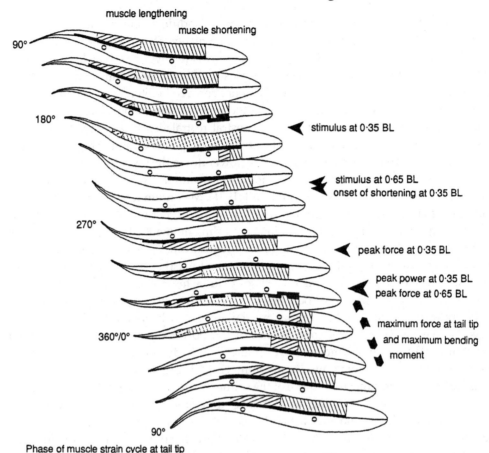

muscle lengthening

muscle shortening

90°

180°

270°

360°/0°

90°

stimulus at 0·35 BL

stimulus at 0·65 BL
onset of shortening at 0·35 BL

peak force at 0·35 BL

peak power at 0·35 BL
peak force at 0·65 BL

maximum force at tail tip
and maximum bending
moment

Phase of muscle strain cycle at tail tip

Figure 7. A schematic representation of the events of a single tail beat, showing progressive fish outlines and midlines from above, in 30° increments (fish outlines redrawn from Videler & Hess, 1984, with permission). For a detailed explanation see the text. (Based on Altringham *et al.*, 1993.)

van Leeuwen *et al.*, 1990; Wardle & Videler, 1993), there is a change in muscle function along the length of the body. Rostrally, active muscle shortens and generates power; caudally, active muscle stiffens, resisting stretch, for the transmission of this power to the tail-blade. A transition zone, in which the muscle's role switches from power transmitter to power generator, travels caudally during the tail beat. Travelling from head to tail, myotomes spend an increasing proportion of the early phase of the tail beat transmitting power caudally, generating power over progressively later and shorter phases of the cycle, as the tail-blade completes its sweep. However, all but the most caudal myotomes contribute to the movement of the tail with a power-generating phase towards the end of the tail beat.

Maximum tail-beat frequency will be determined largely by the rate at which rostral myotomes can perform work cycles, explaining their short contraction time. The shorter the twitch, the greater the number of positive work cycles per second. Why caudal myotomes should have longer contraction times is less obvious, but here is a tentative explanation. The slower caudal muscle is probably energetically more efficient than the rostral muscle, and because of the rather different conditions under which it operates, does not need to be as fast. Caudal myotomes are switched off early in the cycle, and are active for only a short part of each tail beat relative to rostral myotomes (van Leeuwen et al., 1990; Wardle & Videler, 1993). However, because of their longer contraction time, force generation persists for some time after activation. Force is therefore high late in the cycle and positive work is still performed after the power-transmitting phase. The caudal muscle may operate more efficiently than the rostral: rostral muscle sacrifices efficiency for speed of operation, so that it can generate large positive power output at high frequency.

The hypothesized transmission of muscle power down the body to the tail is consistent with accepted theories on steady, carangiform swimming, which show that thrust is developed largely at the tail-blade (Lighthill, 1971). Maximum force is measured at the tail-blade as it crosses the swimming track (Lighthill, 1971; Tang & Wardle, 1992), and this coincides with the incident of maximum bending moment all along the body (Hess & Videler, 1984). At much the same instant rostral muscle generates its maximum power output, and the caudal muscle its maximum force (Figure 7). Rostral myotomes, as observed, should generate peak force before caudal myotomes because of the inertial forces in the caudal part of the body as it reverses its direction of lateral movement.

In some tuna, part of the caudal musculature has been replaced by tendons (Kishinouye, 1923), a logical consequence of the power transmission role of caudal myotomes. Tunniform swimmers have a more rigid body, and are less manoeuvrable than carangiform swimmers, and the myotomal muscle probably operates under somewhat different conditions from those of the saithe. In life, carangiform fish spend much of their time starting, stopping, accelerating, decelerating, and changing direction. Under these conditions the muscle has different functions from those described above and tendon cannot replace muscle in these roles. Furthermore, the muscle is acting only as a transmission element for part of the tail-beat cycle, and generating power at other times: the myotomal muscle of carangiform fish has a complex multifunctional role.

Kinematic and EMG studies of a range of swimming fish suggest that the mechanisms described in Figure 7 may not apply to all fish. The systematic phase-shift between activation and strain cycles, seen so clearly in carangiform fish such as saithe, mackerel and trout, is not seen in anguilliform swimmers, which generate thrust along most of the length of their body (e.g. Williams et al., 1989). Myotomal muscle is also used differently in non-steady swimming: during kick and glide swimming in carp (van Leeuwen et al., 1990) and fast starts in sculpin (Johnston et al., 1993) all the fast muscle appears to function primarily in the power-generating mode. The differences in EMG patterns between anguilliform and carangiform swimmers are probably related to the diminishing differences in muscle contractile properties along the body seen from saithe (carangiform) through cod to sculpin (subcarangiform) (Altringham et al., 1993; Davies & Johnston, 1993; Johnston et al., 1993). The exact functional significance of these differences remains unknown, but these observations are discussed in more detail by Wardle & Videler (chapter 8).

I would like to thank Ian Johnston for giving me access to unpublished results. This work was supported by the Royal Society.

Chapter 8

The timing of lateral muscle strain and EMG activity in different species of steadily swimming fish

C. S. WARDLE* AND J.J.VIDELER[†]

*SOAFD Marine Laboratory, PO Box 101, Aberdeen, AB9 8DB
[†]Department of Marine Biology, University of Groningen, PO Box 14,
9750 AA, Haren, The Netherlands

Waves of electrical activity, measured as electromyograms (EMGs), running head to tail through the myotomes of the lateral swimming muscle are a common feature of forward steady speed swimming of fish. The detail of the EMG-onset and EMG-end timing has recently been shown to be of great significance when related to the strain (length change) cycle of the myotomal muscle fibres. The strain cycle is caused by the waves of curvature which pass along the body at a slower speed than the EMG waves during steady speed swimming. Based on this recent model, studies of seven fish species with and without tail blades are compared, and differences are indicated in the ways the myotomal muscle is used in different swimming modes.

INTRODUCTION

Historical overview

Pettigrew (1873) studied a swimming sturgeon and showed that at any given time the body simultaneously showed lateral bending to the left and the right giving it an 'S' shape. The dynamic character of fish swimming movements became apparent once recorded by cine film techniques, and Housay (1912) was the first to describe how a wave of curvature runs from head to tail in elongated fish.

We are now aware that the waves of curvature running along the body of steadily swimming fish are the result of the combined effect of the activity patterns generated in

Maddock, L., Bone, Q. & Rayner, J.M.V. (ed.). *Mechanics and Physiology of Animal Swimming.*
© 1994. Cambridge University Press.

the myotomes, and the interaction between the fish's body and reactive forces from the water in which the fish moves. This implies that the coupling between the waves of muscle activity and the waves of curvature depend on the details of the body and fin shapes, which are a characteristic of each fish species. Several studies using electromyography, have shown in quite different fish types that the wave of muscle activity can travel faster than the wave of curvature (Blight, 1976; Grillner & Kashin, 1976). Electrodes placed among the fibres of a muscle pick up action potentials in fibres near the electrode and usefully indicate their activity as an electromyogram (EMG). EMG records showing start and end of muscle activity can be made and linked to film frames that show the wave of curvature.

EMG investigations of the use of myotomal muscle in swimming fish have progressed through a series of different questions. The simplest was to show that electrical activity does alternate from one side of the fish to the other, once during each tail-beat cycle. Bone (1966) demonstrated that only the thin strip of red muscle was active during slow swimming in the dogfish. EMG studies in many different fish species have since shown that the recruitment of muscles progresses from the thin red strip to the bulk of the white muscle as speeds increase from gentle sustained swimming to fast dashes (Rayner & Keenan, 1967; Hudson, 1973; Johnston et al., 1977; Bone et al., 1978; Robert & Graham, 1979; Brill & Dizon, 1979; Freadman, 1979; Tsukamoto, 1981, 1984). Most studies have shown that EMG electrodes placed in the white muscle show no activity during slow sustained swimming. The white muscle electrode does not pick up red muscle signals when more than a few millimetres from the active red tissue (Bone, 1966). Electrodes in red muscle are swamped by the large signals from the larger cross-sectional area of white muscle tissue that is active at high swimming speeds, making it difficult to monitor the role of red muscle during fast swimming (Bone, 1966).

There have now been a number of studies of EMG in steadily swimming fish of different species with quite different body shapes, where several electrodes are used along the length of the lateral muscle. They have all described a wave of electrical activity linked to the wave of curvature of the fish's body, and the differences in the timing of these two waves have been pointed out (Blight, 1976; Grillner & Kashin, 1976; Williams et al., 1989; Leeuwen et al., 1990; Wardle & Videler, 1993). At first glance the results seem to be very diverse, but the significance of the different timings of the two waves, and the small differences in the relative timing among species, are just beginning to be appreciated, as we can now relate these timings to the mode of muscle use. The present chapter looks more carefully at the published records of these timings in different species and makes them more easily compared by presenting them all in one format. This format reveals an ordered variation in the timings among fish species with different swimming styles.

The waves of EMG-onset, curvature and bending moment

High-speed cine pictures of steadily swimming saithe (*Pollachius virens*) and mackerel (*Scomber scombrus*) have been analysed and described as harmonic motions in terms of Fourier series (Videler & Hess, 1984). This kinematic analysis was used in combination with Lighthill's (1960) slender body theory to predict the timing of the waves of the maximum bending moment down the body of saithe and eel (*Anguilla anguilla*) (Hess & Videler, 1984; Hess, 1983). The speed of the waves of maximum bending moment turned out to be much faster than the waves of maximum curvature. The analysis of saithe kinematics suggested an instantaneous maximum bending moment all along the body from head to tail, while the wave of maximum curvature took approximately one period, T, to travel from head to tail. The wave of maximum bending moment of the eel ran down the body 3·75 times as fast as the waves of maximum curvature. The saithe and mackerel have large tail blades, while the eel tapers to a point. These distinct patterns in the behaviour of maximum bending moment confirm the continuous nature of the thrust during eel-like motions, in contrast with the pulsed nature of thrust developed at the tail tip in saithe and mackerel.

Waves of curvature give all phases of muscle strain

When swimming at steady speed, the bodies of saithe and mackerel always contain one complete wave of curvature within their length (Videler & Hess, 1984). Muscle, as well as all other soft tissue, goes through strain (length change) cycles due to the repeated motions of the bending body. At any one position on the body a complete strain cycle covers 360°. To help with discussion of Figure 1, the strain in the left lateral muscle, at any point showing maximum concave curvature, can be defined as in a fully shortened state, with phase of 270°. In a similar way the left lateral muscle, at resting

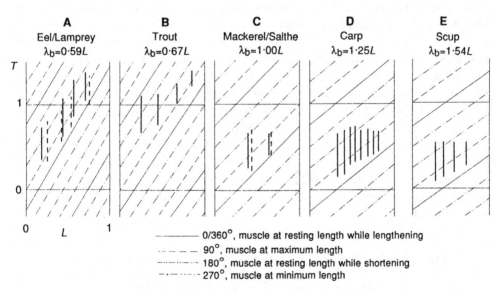

Figure 1. In each of the diagrams the vertical axis indicates time in cycle period, T, and the horizontal axis represents the body length between head ($L=0$) and tail ($L=1$). The vertical bars indicate the duration of the EMGs at that position on left side of the body (EMGs of lamprey and saithe are dashed). The sets of diagonal lines in each diagram represent left side muscle strain phases of the swimming cycle. The horizontal line at $T=0$ marks the start and at $T=1$ the end of one cycle. λ_b is the body curvature wavelength given in terms of body length (L). The five diagrams are ordered with respect to λ_b. See text for details.

length while extending, has phase 0/360°, left fully extended or convex has phase 90°, and left at resting length while shortening, 180°. These phases are as defined by Altringham & Johnston (1990b). As we shall see, the kinematics of other fish species show that during steady speed swimming the body can contain more or less than one wave of curvature. In the data collected in Figure 1, one wave of curvature in the eel occupies 0·59L, whereas in the mackerel and saithe it is 1·0L, and in the scup (*Stenotomus chrysops*) 1·54L, where L is fish body length. This figure shows for seven species the iso-strain lines for the four phases defined above, where T is the period for one complete cycle of strain at any point on the body, and the horizontal scale (L) represents the distance between the tip of the nose (where L=0) to tip of the tail (where L=1). A horizontal line placed at any T shows the distribution of strain phases along the body at that time in the tail-beat cycle. The tail tip is fully left where the 90° phase line meets the right-hand edge of each diagram, and in mackerel and saithe (Figure 1C), with one wave of curvature within the body length, this is at T=0 and T=1.

Muscle works in different modes

As discussed above, in steadily swimming fish it had been noted that the waves of maximum curvature moved less quickly along the body than the waves of EMG-onset, and they are therefore out of phase; this phase difference establishes the function of the muscle at each position along the body as the swimming cycle progresses (Leeuwen *et al.*, 1990; Altringham *et al.*, 1993).

The relationship between muscle strain and muscle stimulus has been studied in isolated myotomal swimming muscle. The length of muscle fibres is continuously changing in a swimming fish. The timing, duration and frequency of stimulation determine to a large extent the output of the muscles in terms of force, work and power (see Videler, 1993, for an introduction). Isolated bundles of muscle fibres can be stimulated at different phases of their strain cycle by mounting them in a device that precisely controls the oscillation of the length cycle. By this means the muscle can be set up as if it were in the body of the fish, and the forces and length changes can be monitored to give the amount of work for different combinations of the strain phases and stimulus times found at each position in the fish body (Altringham *et al.*, 1993; Altringham, chapter 7). In isolated muscle, the phase of strain at which the stimulus is applied can be optimized for maximum work, power or force output (Altringham & Johnston, 1990a,b).

THE TIMING OF EMG IN RELATION TO BODY MOVEMENTS

Mackerel and saithe, the basis of the comparisons

The kinematic study of saithe and mackerel gives detailed description of the lateral deflection of 11 equidistant points on the body (Videler & Hess, 1984, figure 10). These are used to position the iso-strain lines as the dashed and dotted lines in Figure 1C. The mackerel and saithe EMG timings of EMG-onset to EMG-end are placed as vertical lines on Figure 1C from data in Wardle & Videler (1993, table 1). These two species were trained to swim while trailing long, fine EMG wires back and forth across a tank 8 m wide in response to feeding signal lights (Wardle & Videler, 1993), and the results shown in Figure 1C are based on bouts of swimming selected for steady speed at a wide range of speeds.

In Figure 1C, EMG signals from two points at about 40%L and 65%L in the left lateral muscle of mackerel (*Scomber scombrus*; L=0·28-0·33 m), and saithe (L=0·42-0·50 m), were recorded synchronously with films of steady, straight swimming motions (Wardle & Videler, 1993). In mackerel and saithe respectively, onset of EMG activity at the front was at approximately 45° and 35° in the strain cycle and at 74%T and 77%T before the left-most tail blade position (T=0 or T=1, Figure 1C); rear EMG-onset occurred at 0° phase or 15%T and at 18%T after front EMG-onset. The duration of the EMG burst is longer at the front position (41%T and 47%T) than at the rear (25%T and 27%T). At all swimming speeds, the wave of electrical activation of the muscle travelled between the two electrodes 25%L apart at a velocity between 1·5 and 1·6LT^{-1}. In both species, the duration of EMG activity at both electrodes remains a constant proportion of the tail cycle period, T, at all the tail-beat frequencies between 1·8 and 13 Hz. In mackerel and saithe, at tail-beat frequencies up to 3 Hz, only red muscle is involved, and at frequencies above 4 Hz white muscle becomes active. In practice, this change-over matches the change from sustained swimming to prolonged swimming during endurance speed tests in mackerel and saithe (He & Wardle, 1988).

The five studies of species compared in Figure 1 all use treadmill flume techniques to film and record synchronously EMG through short wires running from the swimming fish. Some of the studies compared below include only a single tail beat. Some have more electrodes and achieve synchronous film of the whole fish body movement. All these approaches have independently given similar results, but, when transferred to the single comparable time/length format in Figure 1, reveal interesting differences that may be significant in showing how the muscle is being used during the tail-beat cycle.

Two species without tail blades

Grillner & Kashin (1976) reported a detailed study of two tail beats of an eel (L=0·411 m), and in their figure 1c they show with dotted lines the times at eight body positions, 13%L to 15%L apart, when the muscle is at resting length while extending (0/360°), and resting length while shortening (180°). The EMG was monitored synchronously using four electrodes on one side. By measuring these EMG timings from their figure 1c and with reference to the same data in their figures 3, 5 and 7 it was found possible to transfer their EMG-onset-end times and muscle strain times to Figure 1A.

The lamprey, *Lampetra fluviatilis*, is another swimmer without a tail blade and an individual 0·35 m long, swimming at 0·7 m s⁻¹, was recorded by Williams *et al.* (1989). They used five points on the body to describe the movements of body sections, and four points were synchronously monitored for EMG timing. Using the same extraction and translation process the data from their figure 1a are plotted together with those from the eel of Grillner & Kashin (1976) on Figure 1A (dashed lines).

The trout

Williams et al. (1989) measured the EMG-onset to EMG-end at four positions, and the bending of the body from film at five positions, in trout (*Oncorhynchus mykiss*; $L=0.28$ m), at a speed of 0.18 m s^{-1}, shown frame by frame in their figure 1. In order to transfer their data to the same format, 'concave' in their figure 1 becomes left muscle fully shortened (phase 270°), and 'convex' becomes left muscle fully extended (phase 90°), and by plotting their data relative to these phases from their figure 1, it can be presented for comparison on Figure 1B.

The carp

Leeuwen et al. (1990) worked with carp (*Cyprinus carpio*; $L=0.156$ m) and fitted eight electrodes spaced along the left lateral muscle. The data, including the electrode spacing used, are extracted from their figure 4 representing one tail cycle of a carp swimming steadily at 0.31 m s^{-1}. To compare the EMG-onset to EMG-end timings, the muscle strain phase $0/360°$ was identified in their figure 4 when successive points of the carp body sequentially cross the centre line going left. These strain lines and the carp EMG-onset to EMG-end are marked on Figure 1D.

The scup

Rome et al. (1993) recorded muscle phase and EMG onset in scup ($L=0.216$ m, Rome, personal communication), swimming steadily at 0.80 m s^{-1}. They used four electrodes on one side at 29, 40, 54 and 70%L and recorded their findings in their table 1. At all these positions on the scup body the EMG-end occurred at the same instant in time. Phase of muscle EMG-onset is expressed in degrees before our phase 90° (fully stretched state). The duty cycle is expressed as percent T between EMG-onset and EMG-end, and is given for each of the body positions. In our Figure 1E (firm vertical lines) we drew the EMG-end times at the same moment at each body position, and the EMG duration in units of %T is derived from the duty cycle at each position. The slope of the 90° phase line relative to the EMG timing is thus established for comparison.

CONCLUSIONS

Figure 1 has been ordered according to the length of the wave of curvature in terms of body length. Similar data sets for eel and lamprey, Figure 1A, and mackerel and saithe, Figure 1C, have been plotted together. In Figure 1C there is one wave of curvature on the body, and the EMG-ends at the two positions occur nearly 90° out of phase at points only 25%L apart. EMG is active until the lateral muscle is half-shortened (180°) at 40%L, but only until it is fully extended (90°) at 65%L.

In order to interpret these effects in a swimming fish, the characteristics of the myotomal muscle tissue were studied by Altringham *et al.* (1993) in isolated live muscle fibre bundles taken from the saithe at these same positions, and these were stimulated using similar timings to the EMG and strain phases shown in Figure 1C. A model developed from these results suggests how work from muscle fibres in rostral myotomes can reach the tail blade to generate thrust by appropriate cycles of synchronized mode changes in muscle fibres at the intermediate positions. For example, at the front position the EMG-onset (Figure 1C, lower end of the vertical bars), at 45° and 35°, should give large positive work output. At the rear position the EMG-onset occurring at 0° phase, while the muscle is still only half extended while being stretched, should generate negative work and high force. Although rostral muscle spends more time doing work and caudal muscles spend more time in a stiffened, high force, tendon-like transmission mode, each fibre at each position can change its function for appropriate periods during the progress of the tail-beat cycle. This aspect of the story is given in greater detail in the recent paper containing the model by Altringham *et al.* (1993). See also Altringham, chapter 7, and Videler (1993) for an overview.

The bodies of the eel and lamprey shown in Figure 1A include 1·7 waves within their length, and each wave is 0·59L long. Note from Figure 1A that the zones sampled for EMG in the eel and lamprey include more than one complete wave of curvature, whereas the zones sampled in the mackerel and saithe, Figure 1C, include only one-quarter of a wave of curvature. In the eel and lamprey the EMG-ends (upper ends of vertical bars) only just cross through 90° of muscle strain phase, and the EMG-bars show a distinct upward trend and are nearly parallel to the iso-strain lines.

In the scup (Figure 1E) the sampled zone extends through 41%L, but there is only 0·65 of a wave within the length of the body, so the sample includes 27% of a wave, a similar proportion to that sampled in saithe and mackerel (Figure 1C). The EMG onset and end at 40%L in scup are at similar strain phases to those of the saithe. At 70%L activity is later than in mackerel and saithe at 65%L. The slightly later EMG-onset and end at this position is significant in maintaining positive work through a major part of the tail-beat cycle close to the tail (Rome *et al.*, 1993). This contrasts with negative work being done at this position in the saithe (Altringham *et al.*, 1993). The trout and carp examples show intermediate states where their waves of curvature are 0·67L and 1·25L respectively, with 1·5 and 0·8 waves of curvature within their body lengths.

The very similar and characteristic timings of the eel and lamprey systems (Figure 1A) can be related to the fact that the work done by the muscle is passed as thrust directly and continuously from the muscle to the water. This mechanism is very distinct from fish species with a streamlined, rounded body, narrow tail peduncle and a well-formed tail blade, as found in saithe and mackerel. From the model of Altringham *et al.* (1993), the saithe is organized in such a way as to transmit a major part of the force generated in the thicker rostral myotomes via caudal myotomes to the tail blade, where it is passed as a pulse of thrust to the water. The scup is interesting in presenting something of the eel's characteristics in its muscle phase and EMG timing, and it also has a high body with high dorsal and ventral fins that might be found to transfer force directly to the water. This style of swimming, with short stride length and short wave length of body curvature, might be a characteristic of other flattened fish, but details are clearly needed of swimming kinematics and EMG timings in more of the diverse forms of fish species.

In conclusion, although there are similarities between the fish studied there are also differences that need further investigation. Studies of such an apparently dynamic system need to be rigorously defined in order to make repeatable and comparable recordings and to find the patterns associated with each type of movement. There are problems in making fish swim naturally, in particular when attached to invasive electrodes. So far there do appear to be patterns that make sense and further careful investigations with well-defined questions and careful techniques could be rewarding.

The role of the muscle fibres of each myotome may be modified depending on their use. Fish with tail blades may be able to choose to swim with eel-like motion by erecting dorsal and ventral fins and applying these surfaces and their body wall to the water by using a slightly modified sequence of bending of the body. In presentations such as Figure 1 this might show up as a slight anti-clockwise rotation of the EMG-onset-end sequence, making it more parallel with the iso-strain lines, thus generating more local work output and less longitudinal transmission. Gray (1933b) recorded the swimming of a whiting without a tail blade, and the frame-by-frame records of the wave of curvature show this to be modified. Fish that have damaged their tail blade and have the trailing edge reduced or missing will automatically, due to loss of the reaction from the tail blade, modify their strain cycle and switch to a more eel-like motion, and the EMG-to-strain timing relationships will show the suggested swing. There are a number of deep-sea rattail fish with saithe-like muscular bodies, but their body tapers to the typical rattail with no tail blade, and they use ventral and dorsal fins to transmit thrust directly to the water in eel-like swimming. Eels and rattails have the advantage of being able to swim rapidly backwards with a reverse body wave. The tail blade of fish like saithe and mackerel makes this impossible.

In all styles of swimming, higher speed involves recruitment of larger cross-sections of lateral myotomal muscle in proportion to the force needed. In the same way, in saithe-like swimming, the stiffness of the caudal muscle is adjusted to be appropriate by muscle recruitment at each speed. We already know that the stiffness of the tail blade is adjustable due to the intrinsic structure of the fin-rays (Videler, 1977).

Chapter 9

Swimming in the lamprey: modelling the neural pattern generation, the body dynamics and the fluid mechanics

J.C. CARLING*, G. BOWTELL[†] AND T.L. WILLIAMS*

*Department of Physiology, St George's Hospital Medical School,
University of London, Cranmer Terrace, Tooting, London, SW17 0RE, UK
[†]Department of Mathematics, City University, Northampton Square,
London, EC1V 0HB, UK

This paper investigates the overall problem of lamprey locomotion. A simple connectionist network is used to model the central pattern generator found within the lamprey spinal cord. The formation of the wave of body curvature produced by neural activity is modelled in terms of a mechanical structure composed of elementary components. A new approach to the problem of aquatic propulsion is proposed in which the standard equations of hydrodynamics are solved numerically for a given motion of the body.

INTRODUCTION

The lamprey is an anguilliform swimmer: thrust is developed by the rostral-caudal passage of a wave of curvature over most of the body length. The patterns of muscle activity giving rise to this curvature are generated by pattern-generating circuitry within the spinal cord and continuously modified by sensory feedback (see Figure 1). Movement arises from the interactions of neural activity, the biochemical and physical properties of muscle, and the mechanical properties of both the body and the water. In this chapter we briefly discuss three aspects of this complex system. A brief review of work on pattern generation by the lamprey spinal cord will be followed by a presentation of a mechanical model of the lamprey body, and its interactions with muscle

Maddock, L., Bone, Q. & Rayner, J.M.V. (ed.). *Mechanics and Physiology of Animal Swimming.*
© 1994. Cambridge University Press.

activation. Finally, the results of some initial studies of the hydrodynamics of anguilliform swimming will be given, along with an outline of future work to be carried out.

CENTRAL GENERATION OF THE LOCOMOTOR PATTERN

The pattern of muscle activation during swimming consists of alternating bursts of activity in the left and right ventral roots of each segment, with a rostral-caudal delay of activation along the length of the spinal cord (Figure 1). This pattern can also be recorded from the ventral roots of the isolated spinal cord *in vitro*, in the presence of excitatory amino acids (Grillner *et al.*, 1991). Thus it is clear that the pattern can be generated centrally, without the need of patterned descending control or sensory feedback. Neural activity such as this, which would give rise to locomotor movements if the muscles were intact, has been termed 'fictive locomotion'.

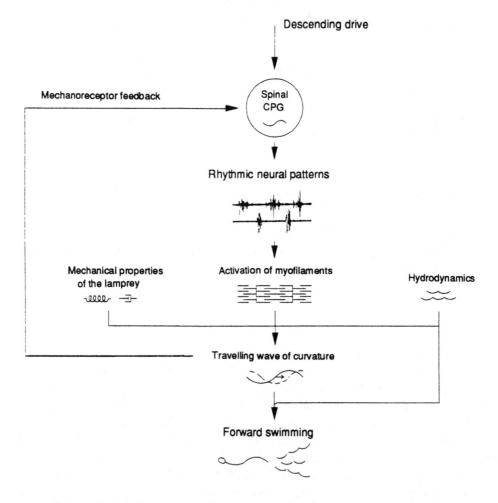

Figure 1. Control of swimming in the lamprey: components which must be considered in understanding how swimming occurs. (Modified from Sigvardt & Williams, 1992.)

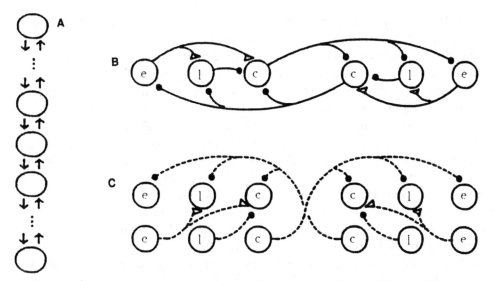

Figure 2. Model of the lamprey spinal cord as a chain of coupled oscillators. (A) Chain of oscillators with nearest-neighbour coupling. (B) Unit oscillator consisting of three cell types: e, excitatory interneuron; l, lateral interneuron; c, crossed caudal interneuron (Buchanan & Grillner, 1987). ●, inhibitory synapses; △, excitatory synapses. (C) Ascending coupling between two oscillators; connections within each segment not shown for clarity. (Modified from Williams, 1992.)

It has been shown that as few as two spinal cord segments can generate the rhythmic pattern and that these segments can be taken from anywhere along the spinal cord, so the lamprey central pattern generator (CPG) can be viewed as a series of unit oscillators (Figure 2A). Faster swimming in the lamprey is accomplished by a decrease in the cycle duration. The intersegmental time delays scale with the cycle duration, so that the delay between segments remains a constant fraction of the cycle at all swimming frequencies, remaining equal to approximately 1% of the cycle per spinal cord segment. Since the lamprey has approximately 100 spinal segments, this ensures that approximately one full wavelength of activation occurs at any instant, at all speeds. This phase-constant feature of the pattern is also seen with fictive locomotion generated by the spinal cord *in vitro* (Wallén & Williams, 1984). This phase delay cannot be simply composed of synaptic or conduction delays between segments, because they would not scale in this way. The intersegmental phase coupling must be an emergent property of the spinal cord circuitry, and for this reason the lamprey spinal cord has been viewed as a chain of coupled oscillators (Cohen *et al.*, 1992; Sigvardt & Williams, 1992).

The detailed structure of the segmental oscillators in the lamprey is not known. It has been postulated, however, that the unit oscillator network is comprised of three interneuron types (two inhibitory and one excitatory) with synaptic connections as shown in Figure 2B (Buchanan & Grillner, 1987). Computer simulations have shown that such a network produces oscillatory patterns, with timing between the neurons similar to that seen in the lamprey (Grillner *et al.*, 1988). Simulations of the lamprey CPG have been performed by synaptically coupling such a unit oscillator into a chain (Williams, 1992). The simulations were performed using a connectionist model based on standard principles of synaptic current flow and membrane potential control, but action potentials were not simulated. Instead, the height of a cell's membrane potential above threshold is taken as a measure of the firing rate of the cell, and the conductance change in a postsynaptic cell is proportional to this firing rate.

The form of coupling between the unit oscillators used in the simulations was based on a simple, biologically plausible, developmental rule: whatever synaptic contacts each neuron makes within its own segment, it must also make in neighbouring segments, but with a smaller synaptic strength (Figure 2C). This coupling has been called 'synaptic spread'. When two equally activated oscillators are coupled by synaptic spread in one direction, the phase relation that develops between them is that for which the sending oscillator neither speeds up nor slows down the receiving oscillator. For this particular form of unit oscillator and coupling, the phase lag that develops is such that the segment receiving the coupling signals leads. Thus in this circuit caudal-to-rostral coupling gives rise to a rostral-to-caudal delay.

A chain of such oscillators, all with the same intrinsic frequency and with coupling in only one direction, shows a uniform intersegmental phase lag. There is much evidence in the lamprey, however, that intersegmental coupling is both rostral and caudal. If the coupling is in both directions with equal strength, then a symmetric pattern results, in which waves of activation travel from the centre of the chain towards both ends. This behaviour is consistent with the mathematical analysis of coupled oscillators (Kopell & Ermentrout, 1988) but is not biologically realistic.

If, however, the intersegmental coupling is of unequal strength in the two directions, the resulting phase lag is uniform along all but a few segments at one end of the chain. In the simulations the ascending coupling was made dominant, for which there is evidence in the lamprey (Williams *et al.*, 1990). Except for a small boundary layer near the rostral end, the intersegmental phase lag is uniform and approximately equal to that seen when only ascending coupling is present. As long as the descending coupling is weaker than the ascending coupling, its only effect is to produce the boundary layer at the rostral end; if it is greater in strength than the ascending coupling, the phase lag is positive (the wave travels caudal-to-rostral) and the boundary layer is at the caudal end. The phase lag is relatively independent of the strength of the coupling. In addition, it was found to be independent of oscillator frequency (Williams, 1992), so this chain of coupled oscillators is appropriate for modelling the CPG of the lamprey.

A MODEL OF ANGUILLIFORM BODY DYNAMICS

In this description the body of the anguilliform swimmer is modelled as a set of dynamical elements (M_s, in Figure 3), each of which is attached via perpendicular extensions of length, w, to the midpoint of a smoothly linked rod of length, l. Because the mass of the animal is symmetrical about its midline the model places equal point masses at each of the pivots, plus one at each of the ends. No other elements are assumed to have mass and the complete structure is assumed to move in only two dimensions. This last constraint is not considered to be too restrictive since during

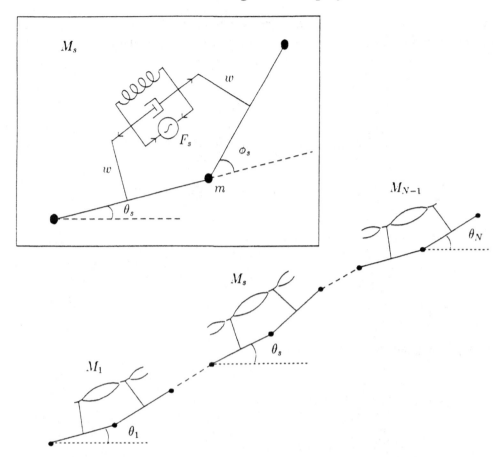

Figure 3. Representation of anguilliform body: muscle segment, M_s, consists of a spring of stiffness μ_s, a dashpot with damping coefficient, γ_s, and forcing term, $Fs(t)$. The N light rods of length, l, are smoothly jointed, with masses, m, placed at each pivot and at both ends; the rods are controlled by arms of length, w, fixed at right angles to each of their midpoints. θ_s and ϕ_s are respectively the angle between rod, s, and the x axis and the angle between adjacent rods in segment, s. (Modified from Bowtell & Williams, 1993.)

swimming there are long periods when such a motion can be observed. In the real animal the swimming muscles are symmetrical about the midline, and activation of muscle on the left and right sides of the body at a given segment will produce turning couples of opposite sign at the corresponding pivot. We have therefore simplified the muscle component in the model to a single-sided element which is capable of producing both positive and negative contractile forces corresponding to the sum of the forces produced by the muscles on both sides of the midline. In detail, each of the elements, M_s, consists of a power unit which is capable of producing a time-dependent thrust or tension, plus a damping element modelled by a linear velocity-dependent dashpot, and a stiffness element modelled by a simple linear spring. The latter two components of damping and elasticity are included to model the total viscous and stiffness properties of all the body tissue; namely the muscle itself as well as the skin, connective tissue and the spinal column or notochord. In terms of the adapted co-ordinates $\{\theta_1,...,\theta_N\}$ and a generalized force, G_{θ_s}, associated with each segment, the equations of motion are given by:

$$
(N-s+1)\sum_{i=1}^{s}\frac{d^2\theta_i}{dt^2}\cos(\theta_i-\theta_s)+\sum_{i=s+1}^{N+1}(N-i+1)\frac{d^2\theta_i}{dt^2}\cos(\theta_s-\theta_i)
$$

$$
+\sum_{i=s+1}^{N+1}(N-i+1)(d\theta_i/dt)^2\sin(\theta_s-\theta_i)-(N-s+1)\sum_{i=1}^{s}(d\theta_i/dt)^2\sin(\theta_i-\theta_s)
$$

$$
+\frac{1}{l}(N-s+1)\left(-\frac{d^2x_o}{dt^2}\sin\theta_s+\frac{d^2y_o}{dt^2}\cos\theta_s\right)
$$

$$
=\frac{1}{ml^2}G_{\theta_s}, \qquad s=1,...,N \tag{1}
$$

and the head position (x_o,y_o) is given by:

$$
x_o=\frac{lN}{2}-\frac{l}{N+1}\sum_{i=1}^{N}(N-i+1)\cos\theta_i,
$$

$$
y_o=-\frac{l}{N+1}\sum_{i=1}^{N}(N-i+1)\sin\theta_i.
$$

For details of the derivation of these equations see Bowtell & Williams (1991).

As well as containing stiffness and damping terms the generalized force G_{θ_s} contains the forcing function $F_s(t)$ which represents the time-course of activation of the contractile protein filaments. Electromyogram recordings from swimming animals show that each burst of ventral root activity occupies approximately 40% of a cycle, with strict left-right alternation (Wallén & Williams, 1984). Because activity on the contralateral side is modelled by using negative values, this pattern can typically be approximated by a sine wave or suitable square wave. Our investigations have included both these types plus the facility to attenuate the activation at each end of the body. It is perhaps worth noting here that all types gave qualitatively similar results. The non-linear nature of the equations of motion suggest an investigation based on a simple linearization. Assuming low curvature dynamics with slow swimming speeds the equations of motion (1) can be linearized and reformulated in terms of the angles ϕ_s between adjacent rods and the angle θ_1 (see Figure 3). This gives:

$$\frac{d^2\theta_1}{dt^2} = -\frac{1}{ml^2}(G_2 - 3G_1)$$

$$\frac{d^2\phi_s}{dt^2} = -\frac{\omega}{ml^2}(G_{s+2} - 4G_{s+1} + 6G_s - 4G_{s-1} + G_{s-2}), \qquad s = 1, ..., N-1, \tag{2}$$

where the term G_s is given by:

$$G_s = \frac{\omega}{l}\left(\mu_s\phi_s + \gamma_s\frac{d\phi_s}{dt}\right) - F_s(t), \qquad s = 1, ..., N-1.$$

The linearized system given by equations (2) was investigated using standard numerical techniques, and in order to observe the movement induced by the different forcing terms a simple time-lapse graphical output was constructed. In all cases the forcing function $F_s(t)$ was chosen so as to produce a wave of muscle activation with a wavelength of approximately one body length and a frequency of two cycles per second, a typical swimming frequency for the lamprey. Both the total mass and length of the creature were normalized to unity and the half-width, w, was chosen by observing that the width of a typical lamprey is about 4% of its body length. In the actual lamprey there are approximately 100 segments controlling the body movement. However, we found in our investigations that using values of N>20 made very little difference to the final results. Thus the investigation reported here was carried out with N=20.

A more difficult task was the selection of the parameters of stiffness and damping, as these are not easily estimated from knowledge of the intact animal. Although the theory allows for the stiffness and damping to vary from segment to segment, in this initial investigation we assumed they were constant. To choose a reasonable range of values for these parameters we have considered the hypothetical behaviour, in an inactive body, of a single end segment, after being mechanically displaced by an external force. This situation is modelled using the above scheme with N=2, $F_s(t)$=0 and θ_1=0. The parameters of stiffness and damping can now be related in a one-to-one manner to two new and more physical parameters, α and T. The parameter T represents the time required for the single segment to return to 10% of its original displacement, and α is a measure related to any oscillation that might ensue when the inert muscle is released. From an intuitive assessment of the property of lamprey muscle, T was taken to be in the range 0·01 to 1 s, a range of two orders of magnitude. For values of the parameter α ≥1 the problem is under- or critically-damped; it seems reasonable therefore to require that α should not be greater than one as this would give rise to oscillatory behaviour in response to a displacement of an end segment. In our investigations α was taken to be in the range 0·01 to 1·0. One final requirement that imposed restrictions on the allowed parameter space was the time it took for the transients in the solution of the complete active problem to die away. We assumed that any point in the parameter space that corresponded to a time lapse of more than 2 s to reach a steady state to be unacceptable.

This information is of course only available after the corresponding equations have been solved.

The results of the investigation did not produce the travelling waves that are observed in the swimming lamprey, but profiles that were more like standing waves with either two or three nodes. Having failed to obtain results that mimicked the intact swimming lamprey, we decided to observe the behaviour of intact lamprey out of water, on a smooth surface. Superimposed tracings from single video frames were constructed and compared with the output of our numerical results (Bowtell & Williams, 1991). The agreement was encouraging, as typically the lamprey out of water moved with a standing-wave-like profile that contained either two or three nodes, and only when a layer of water was placed on the bench did the movement take on the appearance of a travelling wave and the lamprey make forward (albeit slow) progress.

The similarity between the behaviour of the model over a wide parameter space and the behaviour of an intact lamprey on a smooth surface indicates that this simple linear model for the mechanical structure of the lamprey and the interaction between the muscle filaments and the surrounding tissues may be sufficient for further investigations of the control of movement in this animal. When we first began to model the lamprey body as a viscoelastic structure acted upon by a forcing function, we anticipated that the effect of the surrounding water on the movement of the animal might be incorporated within the lumped parameters of elasticity and viscosity (Williams, 1991). It has now become clear that for further progress the model of the lamprey body must be embedded within a model of the fluid mechanics (Bowtell & Williams, 1991).

Although the musculature of the animal consists of a discrete number of segments, a continuum limit of the linearized equations (2) would be of great value for the further investigation of the system. Bowtell & Williams (1993) have carried out such a study and have been able to obtain results for the phase delay between the wave of activation and the wave of curvature of the body. Their results show that there is a slowing down of the mechanical wave with respect to the wave of activation, something that is observed in the swimming lamprey (Williams et al., 1989). It would appear therefore that this effect is in part due to the mechanical structure of the lamprey and may not depend wholly on hydrodynamical forces.

Apart from embedding the model in water, and apart from investigating the effect of the nonlinear terms, there is one important physical property of the animal that has yet to be considered. In the real animal the force generated by the contractile elements is dependent upon the muscle length and its rate of change with respect to time. This can be included in the model by extending the dependence of the forcing term $F_s(t)$ to the variable ϕ_s and its derivative $d\phi_s/dt$. In addition, in the real animal there is sensory feedback to the neurally generated pattern, which is dependent upon the time course of local curvature. This may be included in the model as a local influence on the activation frequency.

THE HYDRODYNAMICS OF ANGUILLIFORM SWIMMING:
A SQUARE-WAVE APPROXIMATION

The detailed computation of the hydrodynamics of swimming creatures is a complex task, requiring substantial computational overheads. Although there is much work of an experimental or analytical kind, to the authors' knowledge there is little or no published work involving numerical studies using the techniques of computational fluid dynamics. The early experimental work of Gray (1933a) provides much useful data from photographic evidence, in particular that from the swimming of the young eel. The extensive work of Lighthill is well-known; for example, Lighthill (1969) reviews the analytical approach to the hydrodynamics of aquatic propulsion, covering carangiform as well as anguilliform swimming. More recently, a volume edited by Webb & Weihs (1983) and also Blake (1983a) have reviewed both the hydrodynamic aspects and the work on muscle mechanics. An abstract outlining some of the present work has appeared (Carling & Williams, 1991) along with others describing work reported at the Plymouth meeting on which this volume is based.

In setting up a full computational model of anguilliform swimming it is best to approach the task in stages. Thus, although a longer-term aim is to combine the body mechanics and muscle activation work described in previous sections with that of the hydrodynamics, it is sensible to start with a prescribed motion, thereby separating, in the first instance, the computation of the flow outside the body from the dynamics within. Later, based on this experience, a fully interacting body and fluid mechanics model can be established which will predict anguilliform motion as an outcome rather than assume it as a precondition. In this spirit, we report here some early work based on a prescribed motion, a travelling square-wave, which is not the most obvious prescription for anguilliform swimming, but nevertheless has an important role in the development of more comprehensive and realistic simulations. So, as a first approximation to anguilliform swimming, a model is proposed in which a backward-travelling square-wave is used to mirror the function of the wave of curvature along the body. The reason for starting with a square-wave, which is clearly an unrealistic body motion, is to gain an understanding of the fluid dynamical regime in an idealised yet representative geometry. For example, for similar Reynolds numbers, the vortex-shedding pattern for the square-wave case may be similar to that in real life. More importantly, the numerical techniques developed during the implementation of the square-wave model can inform subsequent work on a more realistic anguilliform motion.

An impression of the square-wave motion is seen in Figure 4. Concentrating for the moment on the motion of the body alone, Figure 4A shows the configuration of the hypothetical creature at the start of the fourth complete cycle of its motion. This is frame number 96, since there are 32 frames, or time-steps, per complete cycle. In Figure 4B it can be seen that the square-wave front has moved backwards (to the right in Figure 4) whilst the body of the creature has made its way along a square-wave path so that overall there is an advance to the left. This pattern of behaviour continues, and in Figure 4C, one-half cycle on from Figure 4A, an advance (to the left) has occurred which, in this example, is equal to the amplitude of the square-wave. During this time (half a cycle)

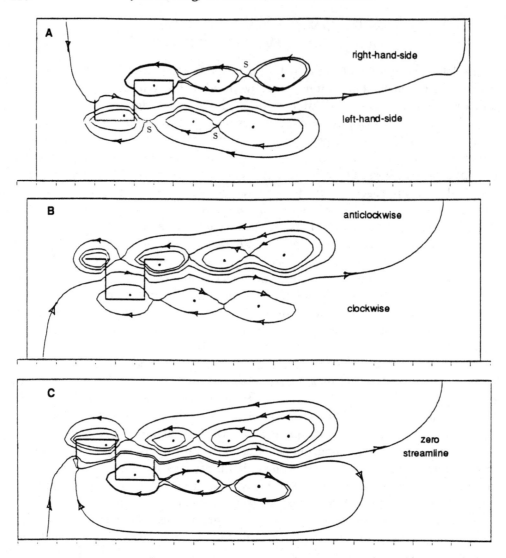

Figure 4. Vortex shedding behind a square-wave creature. Separating streamlines are shown at time frames (A) 96, (B) 104 and (C) 112. Arrowheads indicate direction of circulation.

the backward-travelling wave has also moved (to the right) a distance of one amplitude. The overall advance can be adjusted by varying the ratio of the body speed to the backward-travelling wave speed, where the body speed is strictly along the creature's axis, that is, along the square-wave. The body shape at other points during the cyclic motion can be seen in Figure 5, in which the abscissa shows the frame number and corresponding configuration of the body during the first 1·25 cycles.

The solution of the equations of fluid mechanics for the square-wave moving-boundary problem described above, is best achieved by a simple co-ordinate transformation. For cartesian co-ordinates $\hat{x}$, $\hat{y}$ and $\hat{t}$ where $\hat{x}$ is in the direction of the

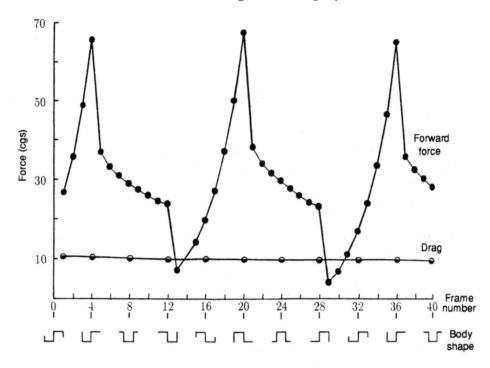

Figure 5. Net forward force and overall drag against frame number (or time-step).

travelling wave (to the right), $\hat{y}$ is directed up the page and $\hat{t}$ represents time, the transformation is:

$$x = \hat{x} - \hat{t}, \quad y = \hat{y}, \quad t = \hat{t}, \tag{3}$$

where the dimensionless speed of the travelling wave is unity. The great advantage in using this transformation is that the path followed by the square-wave, which in real space is undergoing a translation, is now converted into a stationary path in the transformed co-ordinates. Although the computation in (x, y, t) will remain time-dependent, the body of the creature will follow a fixed castellated path in the new co-ordinates. A fixed computational grid can therefore be used in (x, y, t) to represent a moving boundary problem in $(\hat{x}, \hat{y}, \hat{t})$

The discussion above implies that the fluid mechanics is taken to be two-dimensional in the space co-ordinates. It should be pointed out that in reality the fluid flow in anguilliform swimming will always be three-dimensional in character, even for body motion of a strictly planar kind. Before developing the more realistic three-dimensional models, however, it is necessary to investigate the problems, both computational and fluid dynamical, posed by the simpler two-dimensional case. In addition to the two-dimensional (but time-dependent) assumption, the fluid is assumed to be incompressible and isoviscous at intermediate Reynolds number. The Reynolds number, Re, is defined as

$$Re = \frac{da_o\rho}{\mu},$$ (4)

where d is the amplitude of the travelling wave, a_o its speed, ρ the density of water and μ its dynamic viscosity. As an indication of the fluid dynamical regime, based on equation (4), the Reynolds number for the young eel filmed by Gray (1933a), which is approximately 8 cm in length, would be of order 250. On the other hand, for a 30-cm lamprey the Reynolds number might be 100 times greater. The Reynolds number used here for the square-wave creature is 250. On this assumption and using other appropriate parameters to define its movement, the model should give a reasonable approximation to the flow conditions found in the swimming behaviour of the young eel.

The well-known Navier-Stokes equations can be used to describe the fluid mechanics under the conditions defined above. After transformation using equation (3), the governing equations become:

$$\frac{\partial u}{\partial t} + (u-1)\frac{\partial u}{\partial x} + v\frac{\partial u}{\partial y} = -\frac{\partial p}{\partial x} + \frac{1}{Re}\nabla^2 u,$$ (5)

$$\frac{\partial v}{\partial t} + (u-1)\frac{\partial v}{\partial x} + v\frac{\partial v}{\partial y} = -\frac{\partial p}{\partial y} + \frac{1}{Re}\nabla^2 v,$$ (6)

where

$$\nabla^2 \equiv \frac{\partial^2}{\partial x^2} + \frac{\partial^2}{\partial y^2}.$$

The continuity equation is

$$\frac{\partial u}{\partial x} + \frac{\partial v}{\partial y} = 0.$$ (7)

The dependent and independent variables in equations (5)-(7) are dimensionless and have been defined in a manner consistent with the definition of Reynolds number in equation (4). The presence of the factor $(u-1)$ in the formation of the inertia terms in equations (5) and (6) comes from applying the transformation given in equation (3).

In the geometry defined by the transformed square-wave described above, and with appropriate boundary conditions on the velocity components, equations (5), (6) and (7) can be solved numerically at each time-step to give grid values for pressure, p, and the two velocity components, u and v. From the velocity components which satisfy the two-dimensional continuity equation (7), it is possible to derive, by an appropriate integration, the instantaneous stream function. At any instant, therefore, contours of stream function, or stream-lines, can be usefully used to give an impression of the flow field. The volume flow between two given streamlines is always the same, thus giving some sense of discrimination between regions of high and low fluid velocity. The grid size used here is square with side length equal to 0·25 of the square-wave amplitude. As mentioned above, there are 32 time-steps per cycle of the square-wave motion. 'No-slip' conditions have been used along the body of the creature, thus introducing effectively a boundary layer around the square-wave creature as it progresses. This enables viscous drag forces to be calculated. A manuscript describing the numerical technique used to solve the equations is in preparation.

Numerical solutions of time-dependent fluid mechanics tend to generate a great amount of data (even in two-dimensions) and it is difficult to give an impression of the evolution of the flow field in a publication. Nevertheless Figure 4 attempts to illustrate the structure of vortex shedding which occurs behind the square-wave creature. Each part of Figure 4 is taken from a computer-animated film sequence. Overall the film sequence starts from rest and follows the growth and shedding of the vortices through four complete cycles of the square-wave motion. In Figure 4A, at the start of the fourth cycle, it can be seen that, on the right-hand side, there are three vortices, anticlockwise in sense, with one attached to and still being generated by the body whilst two have become detached but remain as distinct rotating structures. On the left-hand side, a similar structure can be seen, although here the sense of rotation is clockwise in all cases and furthermore the two detached vortices are clearly enclosed within an outer enve-lope. This envelope originates from the stagnation point directly behind the leading (attached) vortex which is being generated by the body. The right-hand, anticlockwise vortex structure is separated from the left-hand, clockwise structure by the zero stream-line which follows the course of a fast-moving stream of fluid being forced away from the creature by the backward-travelling wave. To complete the description of the information contained in the stream-function contour plots, the solid circles mark the centres of circulation within each vortex and the contours themselves, the separating streamlines, have been identified by finding the viscous stagnation points (marked 's' in Figure 4A) which must occur between vortices rotating in the same sense. The stag-gered nature of the trailing vortex wake should also be noted. As the creature continues to make progress through its cycle, further vortices are generated. For example, in Figure 4B, eight time-steps later, there are four right-hand-side vortices, two attached and two detached, with each attached vortex possessing its own envelope surrounding the vortices formed downstream. By the time half a cycle has been completed, see Figure 4C, three right-hand vortices have been shed whilst one remains attached to the body. Meanwhile, on the left-hand side, the emergence of a new vortex at the nose of the creature can be picked out. This is accompanied by a streamline which can be seen to envelop all three established left-hand-side vortices. Thus the pattern of growth and shedding begins again.

Whilst it is interesting to gain an understanding of the detailed structure of the flow field, perhaps of greater importance in the study of aquatic propulsion is an analysis of the forces produced by the fluid acting on the body of the creature. Before presenting such results, however, it must be stressed that the present hypothetical creature has been constrained to move along a square-wave path. The forces generated may well be greater than those found in a real creature of similar size. However, the same is true, qualitatively at least, for any prescribed motion, and so the following results should not be dismissed, but rather the limitations should be understood. Bearing this in mind, Figure 5 shows the net forward force exerted by the fluid on the body of the creature, plotted against frame number, or time-step, during the first cycle of motion. This force has been obtained by taking the fluid pressure difference across the body and integrating along all parts of the body which are perpendicular to the overall direction of progress. Also shown in Figure 5 is the drag on the body as a whole (both sides of all parts of the body) calculated from the fluid viscosity and appropriate velocity gradients. The most interesting feature of Figure 5 is not the dramatic semi-cyclic peaks in the forward force but rather the fact that at four points in the cycle the forward force and the drag balance one another. In real 'steady' swimming the forces integrated over a cycle should be zero. Thus it is encouraging that in the present case the balance of forces is partly negative and partly positive during the cycle as a whole. Furthermore the closest balance of forces appears to occur where the square-wave configuration most resembles a real anguilliform shape, for example, frames 28-32 in Figure 5.

Although the results from the square-wave model have proved interesting in their own right, the next stage in the simulation of the hydrodynamics of anguilliform swimming is currently nearing completion. A numerical technique based on local co-ordinate transformations has been used to model a more realistic, but still prescribed, anguilliform motion. It is important to stress, however, that in extending the work in this way much of the numerical and computational detail remains the same as for the square-wave case. Thus the square-wave approximation to anguilliform swimming has proved to be a valuable test-bed for developing an understanding of the hydrodynamics.

CONCLUDING REMARK

When the fluid dynamical investigation of the more realistic form of the body has been completed, the equations of the lamprey body will be coupled with those of the fluid dynamics. At this stage the motion of the body will no longer be prescribed, but will be an outcome of a balance of forces within the system. The output of the simulation of the CPG can then be used as input to the combined model. The overall aim is to develop a simulation in which activation of the spinal cord circuitry will lead to realistic forward swimming of the lamprey.

The authors are grateful to the Science and Engineering Research Council for financial support.

Chapter 10

Swimming capabilities of Mesozoic marine reptiles: a review

JUDY A. MASSARE

Department of Earth Sciences, SUNY College at Brockport,
Brockport, NY 14420, USA and Rochester Institute of Vertebrate
Paleontology, 928 Whalen Road, Penfield, NY 14526, USA

Mesozoic marine reptiles displayed a range of swimming abilities. Jurassic and Cretaceous ichthyosaurs, with their deep, streamlined body, narrow caudal peduncle, and
lunate tail, were probably efficient, fast sustained swimmers. The elongated bodies and
long, broad tails of the early ichthyosaurs and mosasaurs (and marine crocodiles) suggest
that they were adapted for rapid acceleration and were probably ambush predators. More
problematic are the plesiosaurs, which utilized subaqueous flight using two pairs of wing-
shaped appendages. The long-necked plesiosauroids were probably slow swimmers and
relied on a sneak attack to capture prey. The pliosauroids, with shorter necks, more
compact bodies, and larger, broader limbs were probably faster swimmers than the
plesiosauroids.

By estimating the total drag and the amount of energy available for locomotion, relative
sustained swimming capabilities can be estimated for reptiles of the same mass but
different body forms. Calculations suggest that ichthyosaurs were faster than either
plesiosaurs or mosasaurs, and plesiosaurs were slightly faster than mosasaurs. If the
results are scaled to body length rather than mass, the differences in sustained swimming
speeds are even more pronounced. Differences arise from estimates of hydrodynamic
efficiency, shape, and metabolic rate for each kind of reptile, the latter being the most
important factor.

INTRODUCTION

During the Mesozoic Era (245-65 million years ago) the large (>1 m) mobile predators
in marine communities were reptiles. These included the ichthyosaurs (order
Ichthyosauria), plesiosaurs and nothosaurs (order Sauropterygia), marine crocodiles

Maddock, L., Bone, Q. & Rayner, J.M.V. (ed.). *Mechanics and Physiology of Animal Swimming.*

(order Crocodylia: families Teleosauridae and Metriorhynchidae), mosasaurs (order Squamata: family Mosasauridae), placodonts (order Placodontia), and sea turtles (order Chelonia: family Protostegidae) (see Benton, 1990 for general discussion). With the exception of the placodonts and probably the sea turtles, the reptiles preyed upon other swimming animals. Thus understanding their swimming capabilities will shed light on major aspects of their ecology.

Mesozoic marine reptiles exhibited four basic body forms (Figure 1). Bauplan I (post-Triassic ichthyosaurs) was a deep, streamlined body, deepest at the pectoral region and tapering posteriorly to the peduncle of the lunate caudal fin. The limb bones were disc-shaped, except for the short humeri and femora, and formed flippers. Bauplan II (marine crocodiles, mosasaurs, nothosaurs, some Triassic ichthyosaurs) was an elongate, narrow body with a long muscular tail. The proximal limb bones in many showed little modification from those of typical terrestrial forms, although the digits were often elongated. Bauplan III (plesiosaurs) was an ellipsoidal body with two pairs of wing-shaped appendages, and a short tail. The proximal limb bones were elongate but flattened, and the carpals and tarsals were disc-shaped, but the numerous phalanges were more or less cylindrical. Bauplan IV (sea turtles, placodonts) was a dorsoventrally compressed body covered by bony armour. The stocky limbs may have had webbed feet.

Most of these reptiles have no living descendants. Although good fossil skeletons exist, the interpretation of swimming mechanics is often based in large part on analo-

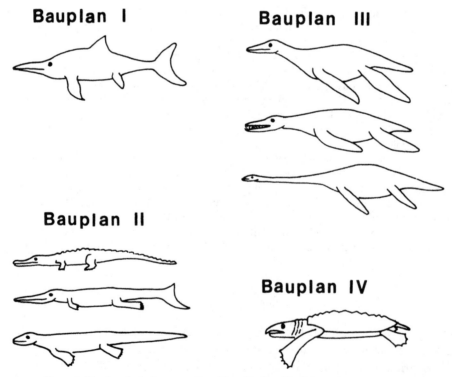

Figure 1. The four body forms, or Baupläne, displayed by Mesozoic marine reptiles.

gies with living aquatic animals. The exact choice of a modern analogue is critical to the picture developed, especially if arguments on swimming abilities are based mainly on analogy. This paper reviews the mechanics of swimming in Baupläne I, II, and III, as exemplified in the ichthyosaurs, mosasaurs, and plesiosaurs, respectively, in an attempt to quantify some differences in swimming among these diverse reptiles.

SWIMMING IN ICHTHYOSAURS

Of all of the Mesozoic marine reptiles, the ichthyosaurs were the most modified from a typical terrestrial reptile. Adults ranged in length from 1 m to 14 m, although most were about 2-3 m long. Their superficial resemblance to living dolphins and thunniform fish, among the fastest, most active swimmers in today's oceans, has led to the assumption that ichthyosaurs were also fast, continuous swimmers. Jurassic ichthyosaurs, the best known and most complete specimens, displayed numerous adaptations for drag reduction necessary for fast sustained swimming. Many had a length/depth ratio of between 4 and 5, the optimum shape for minimizing total drag, and most were within the range 3 to 7 (Massare, 1988). The body was deepest in the pectoral region and narrowest at the caudal peduncle.

The caudal peduncle, although variable in thickness among taxa, was never as narrow as that of a tuna, and was probably more comparable to that of a dolphin. The vertebral column was down-turned at the end, and supported the lower lobe of a lunate caudal fin. In Jurassic forms such as *Stenopterygius* and *Ophthalmosaurus*, the small, simple vertebrae of the caudal fluke were merely a passive support, lacking any musculature (Klima, 1992). In some species the tail fluke was somewhat asymmetric, with the upper lobe slightly smaller (shorter) than the lower (see plates in Hauff, 1953), but in most Jurassic and later species, the fluke was more or less symmetrical (McGowan, 1992). The upper lobe was unsupported by bone, although this does not necessarily mean that it was flexible or passive. The lobes of the caudal fins of whales are not supported by bone, and are quite stiff. Alexander (1989a, p. 126) suggested that the lack of support might be compensated for by a lower aspect ratio to stiffen the tail. The same might be true for ichthyosaurs (McGowan, 1992; Klima, 1992). The aspect ratio of ichthyosaur tails varied among species but was within the range seen in living whales (Table 1; also see McGowan, 1992).

The reversed heterocercal tail (supported lower lobe, unsupported upper lobe) has been explained as an adaptation to neutralize positive buoyancy caused by the presence of lungs (see citations in Taylor, 1987; McGowan, 1983, 1992). Assuming that the tail acted in the opposite sense to the heterocercal tail of a shark, the asymmetry of the thrust generated would produce a downward component that could counteract positive buoyancy caused by the lungs. Taylor (1987) pointed out that the downward thrust component played a role in buoyancy control, but was probably more important for manoeuvring and initiating dives, especially dives from the surface when the animal came up to breathe. The tail could thus be used for manoeuvring as well as propulsion, and the limbs could take advantage of the asymmetric thrust for turning. McGowan

Table 1. *Height/width ratios of some lunate caudal fins. Ratios were calculated from measurements of photographs and drawings to one decimal place. Width was measured as fin width at the caudal peduncle, approximately the maximum fin width.*

Taxon	Height/width ratio	Source
Whales		
Sperm whale	2.5	Minasian *et al.* (1984), p.72
Grey whale	2.5	Minasian *et al.* (1984), p.72
Humpback whale	2.7	Minasian *et al.* (1984), p.72
Baird's beaked whale	3.1	Minasian *et al.* (1984), p. 72
Right whale	3.1	Minasian *et al.* (1984), p.72
Blue whale	3.7	Minasian *et al.* (1984), p.72
Ichthyosaurs		
Stenopterygius macrophasma	3.9	McGowan (1979), pl. 3, fig. 2
S. hauffianus	3.6	Hauff (1953), pl. 7a
S. quadriscissus	2.6	Hauff (1953), pl. 7a
S. quadriscissus (juv)	2.6	Hauff (1953), pl. 9a

(1992), however, challenged the validity of a shark analogy in interpreting the function of the ichthyosaur tail. He argued that the tails of post-Triassic ichthyosaurs were more symmetric and stiffer than the tails of sharks to which they had been compared, and were more like the tails of modern cetaceans or scombroid fish (also see Klima, 1992). The thrust generated probably lacked any significant vertical component, unlike the situation in many sharks with heterocercal tails (McGowan, 1992).

Riess (1984, 1986) challenged the importance of caudal propulsion, arguing instead that many ichthyosaurs swam using primarily their limbs. He pointed out that several specimens supposedly preserving the outline of a caudal fin had been enhanced by preparators who essentially carved out a lunate tail from a carbonaceous film. Furthermore, tail-bends in some ichthyosaurs, notably *Eurhinosaurus*, have been exaggerated by preparators (Riess, 1986). This led Riess to question whether any ichthyosaurs had a down-turned vertebral column or a lunate caudal fin.

The down-turn of the vertebral column is the result of several (three or more) wedge-shaped centra, centra with a slightly greater dorsal length than ventral length. Some Triassic ichthyosaurs may have had straight tails (Appleby, 1979; Massare & Callaway, 1990). Wedge-shaped centra have been reported for only two Triassic genera, *Cymbospondylus* and *Shonisaurus* (Merriam, 1908; Camp, 1980), although the angle of the bend has not been calculated. The type specimen of *Californosaurus* shows a distinct bend in the vertebral column (see plate in Merriam, 1908), but none of the other Triassic genera, including the widespread *Mixosaurus* and *Shastasaurus*, show any evidence of a tail-bend. The lack of a tail-bend, however, does not necessarily mean non-caudal propulsion. *Mixosaurus* has tall neural spines in the mid-caudal region which probably supported some sort of a tail fin. Some Triassic ichthyosaurs had elongated caudal centra that may have served to increase the vertical depth of the tail. Such features have been interpreted as adaptations to increase thrust-production for rapid acceleration in ambush predators (Massare & Callaway, 1990; Massare, 1992).

McGowan (1989, 1990) used computed tomography to measure the dorsal and ventral lengths of wedge-shaped centra and calculate the degree of flexure of the tail-bend in some Jurassic ichthyosaurs. He calculated an angle of about 24° for *Leptopterygius tenuirostris* (McGowan, 1989), and an angle of at least 37° for *Eurhinosaurus* (McGowan, 1990), both based on six centra. The latter genus was a form that Riess (1986) argued lacked a tail-bend. Other estimates of the angle of the tail-bend are based on measurement of the deflection of the vertebral column on preserved skeletons. Such measurements on some Early Jurassic species indicate the following ranges of angles: *Stenopterygius quadriscissus*, 23-39°; *S. hauffianus*, 18-27°; *S. megacephalus*, 21-35°; *S. megalorhinus*, 20-27°; and *S. cuneiceps*, 25-43° (McGowan, 1990, 1992). Although it can be argued that those bends are the results of enhancement by preparators, the range of angles is comparable to those found by measuring the individual centra. The question then arises as to whether the bend does support a lunate caudal fin. Although the accuracy of some specimens is questionable (Riess, 1986), enough other specimens exist showing a lunate tail to verify its relationship to the down-turned vertebral column (McGowan, 1979; Hauff, 1953).

Does a lunate caudal fin necessarily mean that ichthyosaurs swam with their tails? Historically that has been the inference, and it has been supported by recent analyses (Wade, 1984, 1990; Taylor, 1987; Massare, 1988; Klima, 1992; McGowan, 1992). Riess (1984, 1986), however, argued that the similarity in the shoulder girdle, limb shape, and limb proportions of the living Amazon dolphin *Inia* and those of some Jurassic ichthyosaurs (*Eurhinosaurus* and *Stenopterygius*, among others) support the contention that the ichthyosaurs swam with their limbs as does *Inia*. Riess' (1986) analysis, however, has been strongly refuted by Klima (1992), who pointed out that the basis of Riess' argument was a misinterpretation of existing data on the Amazon dolphin. The Amazon dolphin is not an underwater flyer, but uses caudal propulsion as do other cetaceans (Klima, 1992 and citations therein). Reports of limb propulsion for *Inia* may be based on observations of it at slow speeds or while at rest (Wade, 1984, 1990). Klima (1992) further pointed out that the limbs of '*Inia*-type' (*sensu* Riess, 1986) ichthyosaurs are too small and situated too far forward to serve as wings for underwater flight. The argument against underwater flight can be made on ecological grounds as well. The Amazon dolphin inhabits drowned forests for at least half the year (Goulding, 1993; Cater, 1989) where high-speed swimming is limited by the physical complexity of the habitat. Manoeuvring among tree branches does not lend itself to high-speed propulsion. At slow speeds, paddling is the more efficient mode of swimming (Blake, 1980), so it would not be surprising, from biomechanical considerations, if animals in such a habitat replaced caudal propulsion by pectoral propulsion, or at least relied more on their limbs for manoeuvring than did their open water counterparts.

Ichthyosaurs may have used their limbs at slow swimming speeds (Wade, 1984, 1990; Klima, 1992), but caudal propulsion must have been the mode at moderate and high speeds. The large eyes of ichthyosaurs suggest that they lived in clear waters where visibility did not limit swimming speeds, so limb propulsion was probably not important for normal activities. Limbs were probably used for manoeuvring or holding position in the water, as in cetaceans (Wade, 1984, 1990; Klima, 1992; McGowan, 1992). Non-caudal swimming in actinopterygian (ray-finned) fish seems to be secondarily acquired in slow-swimming forms, and substitution of non-caudal propulsion is rare

(Webb, 1982). It seems unlikely that ichthyosaurs as a group would have abandoned caudal propulsion for less efficient (Webb, 1975b) pectoral fin propulsion. There is, however, considerable variation in limb proportions and the amount of tail bend, especially if Triassic ichthyosaurs are considered. The relative importance of the limbs in propulsion, and whether they were used to generate thrust or lift, may likewise have varied. Limb propulsion may have occurred in some atypical species or in unique ecological circumstances, but these remain to be identified.

De Buffrénil & Mazin (1990) examined the microstructure of limb bones of *Ichthyosaurus* (Late Jurassic), *Stenopterygius* (Early Jurassic), and *Omphalosaurus* (Early Triassic). They found that the bones displayed woven-fibred bone tissue, which is considered indicative of a high absolute rate of bone growth; it is absent among living, slow-growing ectotherms. Furthermore, ichthyosaur bone tissue was extensively remodelled with growth, compact bone tissue being transformed into cancellous bone. Buffrénil & Mazin (1990) concluded that ichthyosaurs may have had a high metabolic rate, thus an endothermic or incipiently endothermic physiology. This is not unheard-of in non-mammalian and non-avian taxa: scombroid fish, lamnid sharks and leatherback turtles have high metabolic rates (see citations in Buffrénil & Mazin, 1990). All of them are, in fact, fast sustained swimmers.

Thus, not only did the ichthyosaurs have numerous adaptations for drag reduction, but may have had more power available for locomotion than did the other Mesozoic marine reptiles. Envisaging typical Jurassic ichthyosaurs as fast continuous swimmers, analogous to living dolphins, seems to be a good model. Like living dolphins, they were probably pursuit-predators, continuously active swimmers that searched for prey over wide areas (Massare, 1988).

SWIMMING IN MOSASAURS

Mosasaurs, extinct relatives of monitor lizards, ranged in adult size from about 4 m for *Clidastes* to the 15-m long *Mosasaurus* (Russell, 1967, p. 210). In contrast to the ichthyosaurs, mosasaurs had an elongate shape with a long, broad tail. The mosasaurs differed from other taxa exhibiting Bauplan II in that the proximal limb bones were short and wide. The digits, however, were fairly long and were probably bound in connective tissue to form flipper-like limbs (Mulder & Theunissen, 1986; also see discussion in Lingham-Soliar, 1991). Similarity in body shape to living crocodilians suggests a similar mode of propulsion: a small undulation of the body increases in amplitude towards the rear, resulting in a sweeping undulation of the distal portion of the tail (Manter, 1940; Fish, 1984a). Crocodilians do not undulate the anterior portion of their body; only the pelvic region and tail move during propulsion (Fish, 1984a). Lingham-Soliar (1991) argued, based on the vertebrae construction, that this was the also the case for the largest mosasaurs (e.g. *Tylosaurus, Mosasaurus*); thus their mode of propulsion was axial subundulatory (see terminology in Braun & Reif, 1982). For smaller mosasaurs, however, (e.g. *Clidastes, Platecarpus, Ectenosaurus*), the anterior body was flexible, so axial undulatory propulsion (undulation of the entire body) was likely

(Lingham-Soliar, 1991; Massare, 1988). Mosasaurs probably held their limbs at the side while swimming, or used them for extra thrust during lunges, as is done in living crocodilians (Manter, 1940; Frey, 1982; also see citations in Fish, 1984a).

The tail was a significant portion of the total body length: 43% in *Clidastes*, 50% in *Tylosaurus*, and 55% in *Platecarpus*, the three best-known genera (Russell, 1967, p. 210). With fineness ratios of 7·5 to 12·6, most mosasaurs were well above the optimum ratios for sustained swimming (Massare, 1988). Not surprisingly, mosasaurs displayed many adaptations for rapid acceleration that are characteristic of ambush predators. The elongate shape resulted in a high surface-to-volume ratio for pushing against the water, and a relatively small frontal area for cutting through it. The neural and haemal spines were fairly long throughout the length of the tail, resulting in a broad tail. Two genera, *Clidastes* and *Plotosaurus*, had very elongated neural spines towards the end of the tail. This expanded caudal area may have been an adaptation for increasing thrust production at the distal part of the tail. Lingham-Soliar (1991) suggested that the stiff anterior body of the largest mosasaurs would have maximized ambush efficiency by improving thrust and in focusing on prey.

An interesting insight into mosasaur ecology comes from examining the bone histology of the vertebrae. Rothschild & Martin (1987) and Martin & Rothschild (1989) found evidence of avascular necrosis in two of the three most common genera of North American mosasaurs, *Tylosaurus* and *Platecarpus*. The disease is present in nearly every skeleton, with an average of 25 affected vertebrae per skeleton in *Platecarpus* and nine in *Tylosaurus*. Skeletons of a third genus, *Clidastes*, showed no diseased vertebrae. The authors argued that the most likely cause of the bone disease is decompression sickness, 'the bends'. They suggested that the diving habits of *Tylosaurus* and *Platecarpus* involved either very deep dives or frequent ones. *Clidastes*, on the other hand, was not very likely a diving form. This confirms the suggestions of earlier workers (Williston, 1898; Vaughn & Dawson, 1956) which were based on inferences from morphology.

SWIMMING IN PLESIOSAURS

The plesiosaurs ranged in length from about 1 m to almost 13 m. They relied entirely on their limbs for propulsion. Both pairs of limbs are elongated, tapering structures, and about equal in size and shape. Although forelimb propulsion has been suggested (Brown, 1981), the similarity between the forelimbs and hindlimbs of plesiosaurs suggest that both were used in propulsion. Living aquatic tetrapods that use mainly forelimbs for propulsion (sea turtles, penguins, sea lions) have distinctly different forelimb and hindlimb morphologies.

The observation that the limb girdles of plesiosaurs are expanded in a horizontal plane led to the model of plesiosaurs as subaqueous rowers, moving their limbs mainly in a horizontal plane (Watson, 1924). Robinson (1975), however, presented convincing arguments for underwater flight in plesiosaurs based, in part, on limb shape. The limbs of plesiosaurs taper towards the end, like a wing. Oars, on the other hand, expand distally to increase drag and propel the body forward. Robinson argued that the morphology of the plesiosaur limb was more similar to the wing of a penguin or sea turtle than to the paddle-like leg of a duck or otter (Figure 2). Furthermore, an oar is rotated 90° for the recovery stroke. The oval shape of the glenoid and acetabulum indicates that the plesiosaur limbs could not be rotated 90° for feathering (Nicholls & Russell, 1991).

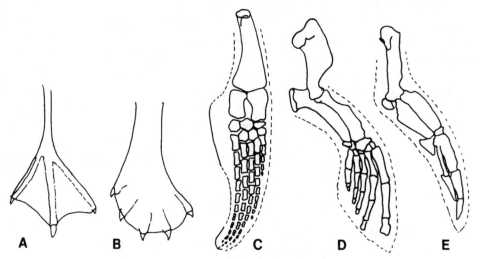

Figure 2. Comparison of limbs of aquatic tetrapods (after Robinson, 1975). (A) Hindlimb of a duck; (B) forelimb of an otter; (C) forelimb of a plesiosaur, *Brachypterygius*; (D) forelimb of a sea lion; (E) penguin wing. (A) and (B) use their limbs as oars; (D) and (E) use their limbs as wings/hydrofoils.

Subaqueous flight requires significant movement in the vertical plane (Figure 3A) The main problem with Robinson's (1975) argument is that the shape of the heads of the propodials indicate that the limbs could not be lifted above the horizontal (Tarsitano & Riess, 1982; Godfrey, 1984; Nicholls & Russell, 1991). Furthermore, the limb girdles are not expanded in the dorsoventral plane. Both the glenoid and the acetabulum are on the same horizontal plane as the girdles, so that no part of the girdle extends ventral to the articular surfaces (Godfrey, 1984; Nicholls & Russell, 1991). The stroke thus could not have been true flight in the sense of penguins because of the lack of a significant vertical component (see Nicholls & Russell, 1991, and Storrs, 1993, for historical review of the flight vs rowing controversy).

Godfrey (1984) likened plesiosaur propulsion to that of an otariid pinneped (sea lion), with a backward and downward power stroke and a forward and upward recovery stroke (Figure 3B,C). The limb acts as a hydrofoil/wing, but the stroke is predominantly a backward and forward motion (Figure 3C) and it is asymmetric in its power output. Godfrey's (1984) model accounts for the limited movement of the propodials, the horizontal expansion of the limb girdles, and the shape of the limbs, while permitting a modified subaqueous flight which is more efficient than rowing (Robinson, 1975; Taylor, 1981). Thus a sea lion would seem to be a reasonable modern analogue for a plesiosaur.

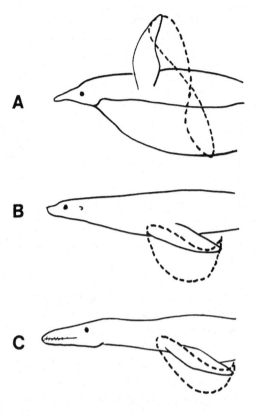

Figure 3. Comparison of subaqueous flight strokes (after Godfrey, 1984). (A) penguin; (B) sea lion; (C) plesiosaur.

Plesiosaur limb girdles and gastralia (ventral ribs) were suspended by muscles and ligaments from the arched vertebral column like the string on an archer's bow (Taylor, 1981; Robinson, 1977). The elastic suspension could absorb some of the limb movement without flexing the vertebral column. Thus plesiosaurs must have had a fairly rigid thorax, a requirement for underwater flight. Plesiosaurs utilized gastroliths (stomach stones) as do most extant underwater flyers, to control buoyancy (Taylor, 1991, 1993, chapter 11).

The old idea of subaqueous rowing assumed that the two pairs of wings would move simultaneously. Would underwater flyers with an asymmetric power stroke necessarily beat their four limbs in unison? Frey & Riess (1982) and Riess & Frey (1991) argued that the two pairs of wings were out of phase by half a stroke, such that the forelimbs were in the power phase (backward, downward movement) while the hindlimbs were in the recovery phase (forward, upward), and vice versa. Thus the plesiosaur was always generating thrust. This would be important to a high drag animal, such as one with a very long neck, in order to maintain a steady swimming speed. The high drag might slow the animal too much during a single recovery phase.

Alexander (1989a) pointed out, however, that swimming would be very inefficient with the fore and hind limbs out of phase because the hindlimbs would be accelerating water that was already in motion. It would be more efficient for the hindlimbs and forelimbs to work together and simultaneously accelerate a larger mass of water. Halstead (1989) also argued for synchronous limb motion, and suggested that plesiosaurs, like most marine tetrapods, did not follow a straight path in the water, but incorporated a component of vertical motion resulting in an undulating or wavy path. His point was that there was no need to adjust the phase difference between the limbs to prevent turbulence generated by the forelimb from increasing the drag on the hindlimb if the animals followed the typical undulating path rather than a straight one (Halstead, 1989). If he is correct, then the limbs must have operated in unison: if they were out of phase, then the vertical component of motion of each limb would be working in opposite directions, an inefficient and energetically costly way to swim. Animals often incorporate a glide phase into their propulsion to save energy. With the forelimbs and hindlimbs operating in unison, the more compact, streamlined kinds of plesiosaurs would have had this option.

On the other hand, Taylor (op. cit.) suggested that there may not have been a fixed phase difference between the two pairs of limbs. In much the same way that a horse can change its gait, plesiosaurs may have adjusted the phase difference between fore and hindlimbs to meet the speed or acceleration requirements of the moment, such that the hindlimb was always in an optimum position to meet the vortex produced by the forelimb (M.A. Taylor, personal communication, 1993). A detailed biomechanical analysis is needed to resolve the question of the phase relationship of the two pairs of limbs.

Within the suborder Plesiosauria, two distinct lineages have been recognized, the plesiosauroids and the pliosauroids, although the monophyly of these two groups has yet to be established. Nevertheless, as historically defined, these groups include two different predator types.

The plesiosauroids had long necks and small heads on a dorsoventrally compressed, ellipsoidal body. Their limbs were fairly slender, and smaller compared to their body length than those of the pliosauroids. There was considerable variation in the length of the neck, reflected in a range of fineness ratios from 5·0 to 13·2 (Massare, 1988). The long neck of some genera must have added drag and slowed them considerably. Alexander (1989a, p. 137) points out that with the neck stretched out in front, plesiosauroids must have had trouble controlling direction when swimming because it put a high surface area in front of the centre of mass. Any movement of the neck would move the plesiosauroid from a straight course. Problems with steering could have been compensated for by a slow swimming speed or by staying at the surface. Plesiosauroid eyes, however, are oriented upward and forward, suggesting that they attacked prey from below. Snapping at fish from the surface would have been difficult, but attacking a school of fish from below was a possibility. The plesiosauroid's small skull would reach the school before the bulkier body could be detected, giving it a distinct advantage for a surprise attack. It would also confer a selective advantage to a longer neck.

The pliosauroids had larger heads and shorter necks, hence overall a more compact body. They were probably more efficient swimmers than the plesiosauroids, lacking a long neck to create steering problems and increase drag. Furthermore, their larger, broader limbs could accelerate larger masses of water per stroke than the generally smaller, more slender limbs of the plesiosauroids. Hence they were probably faster swimmers. Many pliosauroids had fineness ratios within the range for optimum swimming efficiency, so they were probably pursuit predators (Massare, 1988).

HOW DO THESE REPTILES COMPARE?

Can we quantify swimming capabilities so that these animals can be compared directly? Massare (1988) presented an analysis that permits general comparisons among marine reptile groups. For this analysis, eight animals will be considered: two ichthyosaurs, two plesiosauroids, two pliosauroids, and two mosasaurs. Skeletal reconstructions for most of them are shown in Figure 4.

An animal swimming at a constant speed has to counteract the drag by exerting thrust in order to continue at a constant speed. Over the entire propulsive stroke, the two forces are equal. If power output (P) is considered, rather than thrust, then:

$$D\,U = e\,P \tag{1}$$

where D is the total drag on the animal moving at a velocity, U, and e is the proportion of power output used for locomotion (Weihs, 1977; Wu, 1977).

Total drag (D) is given by:

$$D = (\rho\,U^2\,A\,C_d)\,/\,2 \tag{2}$$

where A is the characteristic area (measured as the surface area in this analysis), C_d is the drag coefficient, and ρ is the density of water. Thus as the swimming speed increases, so does the drag. Since there is an upper limit to the amount of power available for locomotion, there is a maximum sustained swimming speed that can be attained. Beyond that speed, the swimming animal is no longer within the range of aerobic exercise, and an oxygen debt is being accrued such that the speed cannot be sustained.

By combining eqn 1 and 2, and solving for U, an expression is obtained for estimating the swimming speed U:

$$U = [\,2\,e\,P\,/\,(\rho\,A\,C_d)\,]^{\,0.3333} \tag{3}$$

If P is measured as the maximum aerobic metabolic rate, then U will be the maximum aerobic (sustained) swimming speed.

Consider the case of animals with different Baupläne, and therefore different fineness ratios, but of equal mass. How would their maximum aerobic swimming speeds compare?

The total power output (P), can be approximated by the metabolic rate, which is given by:

$$P = a\,M^{0.75} \tag{4}$$

(Kleiber, 1947), where M is the mass and a is a constant determined by the level of activity and whether the animal is homoeothermic or poikilothermic. For animals of equal mass, P is proportional to the metabolic constant a.

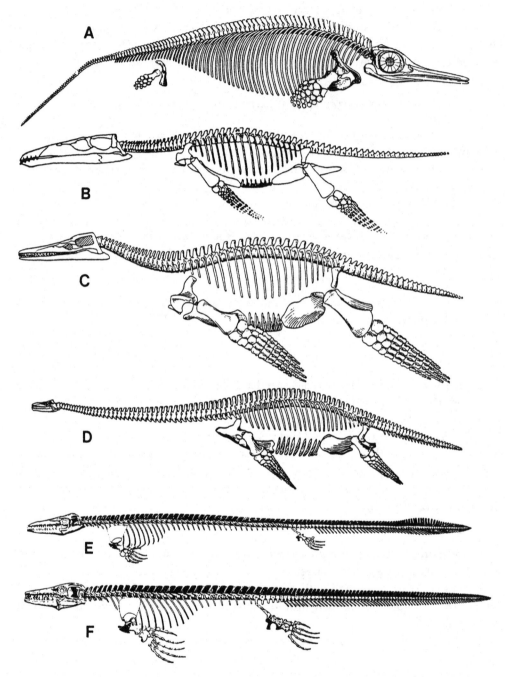

Figure 4. Skeletal reconstructions of some of the marine reptiles used in the analysis of swimming. (A) Ichthyosaur, *Ophthalmosaurus* (after Andrews, 1910); (B) pliosauroid, *Liopleurodon* (after Newman & Tarlo, from Robinson, 1975); (C) pliosauroid, *Peloneustes* (after Andrews, 1913); (D) plesiosauroid, *Muraenosaurus* (after Andrews, 1913); (E) mosasaur, *Platecarpus* (after Williston, from Russell, 1967); (F) mosasaur, *Clidastes* (after Williston, from Russell, 1967).

When the power output is measured as the metabolic rate, the proportion of energy available for locomotion can be expressed as:

$$e = e_p \, e_m \tag{5}$$

where e_p is the propulsive or hydrodynamic efficiency and e_m is the muscle efficiency, the efficiency of converting biochemical energy to mechanical energy (Wu, 1977; Weihs, 1977). If we assume that the muscle efficiency is about the same for all reptiles, then e is proportional to the hydrodynamic efficiency.

The characteristic area, A, is the surface area of the swimming animal. This is difficult to measure in living animals and impossible to measure on extinct ones! For this analysis, the shape is approximated by a prolate spheroid of the appropriate fineness ratio, with the major axis equal to the length of the animal. The characteristic area is therefore the surface area of the prolate spheroid.

Drag on an animal moving through a fluid is the sum of the friction (viscous) drag and pressure drag. For a medium size marine reptile (4 m long), swimming fairly slowly (1 m s^{-1}), the Reynolds number would be approximately 4×10^6, in the range where the pressure drag component is much larger than the friction drag component. By comparison, the pressure drag is 87% of the total drag at a Reynolds number of 10^3, and 97% of the total drag at a Reynolds number of 10^4 for a circular cylinder (Vogel, 1981, p. 76). Furthermore, the transition from laminar to turbulent flow occurs between 5×10^5 and 1×10^7 (Vogel, 1981, pp. 113-114), depending on the exact circumstances. Swimming in most marine reptiles was within this transition range. The drag coefficient for a prolate spheroid in the transition zone can be approximated by:

$$C_d = C_f [\, 1 + 1 \cdot 5 \, (W/L)^{1 \cdot 5} + 7 \, (W/L)^3] \tag{6}$$

where L and W are the lengths of the major and minor axes of the prolate spheroid, and C_f is approximately $0 \cdot 004$ (Hoerner, 1965, pp. 6-17, eq. 8 and pp. 3-12). This neglects drag due to surface waves, but that would probably be similar for all the reptiles considered.

Thus for animals of the same mass, the maximum aerobic swimming speed is proportional to the cube roots of the metabolic constant and hydrodynamic efficiency, and inversely proportional to the cube root of a shape factor.

$$U \propto [\, a \, e_h / (\, A \, C_f (1 + 1 \cdot 5 \, (\, W/ \, L \,)^{1 \cdot 5} + 7 \, (\, W/ \, L \,)^3))]^{\, 0 \cdot 3333}. \tag{7}$$

Table 2 gives the approximate lengths of several marine reptiles of mass 2000 kg. The fineness ratios were measured from reconstructions of the skeletons, neglecting the snout (Massare, 1988). The length was calculated as the major axis of a prolate spheroid with a volume of 2 m^3 and the appropriate fineness ratio. The effect of each of the factors in eqn 7 above can then be examined.

With the power output measured as the maximum aerobic metabolic rate, in J s^{-1}, the metabolic constant, a, for reptiles is in the range of $11 \cdot 6$-$29 \cdot 0$ (converted from metabolic intensities in Bakker, 1975, table 21.1). With a lack of any physiological data, an average constant of $20 \cdot 3$ can be used for the mosasaurs and plesiosaurs. The ichthyosaurs, however, are at least at the upper end of the reptile range, probably even higher (Buffrénil & Mazin, 1990). A constant of $29 \cdot 0$ will be used for ichthyosaurs.

Table 2. *Taxa of marine reptiles, scaled to equivalent mass of 2000 kg. The body shape is approximated by a prolate spheroid with a volume of 2 m³ and the appropriate fineness ratio. Surface area is in m².*

Genus	Bauplan	Fineness ratio	Length (M)	Surface area	Drag coefficient
Ophthalmosaurus	I	3.8	1.9	2.42	0.0053
Ichthyosaurus	I	4.0	2.0	2.45	0.0052
Clidastes	II	12.6	4.2	3.52	0.0041
Platecarpus	II	10.6	3.8	3.32	0.0042
Peloneustes	III	4.9	2.3	2.61	0.0048
Liopleurodon	III	4.6	2.2	2.56	0.0049
Muraenosaurus	III	6.7	2.8	2.87	0.0044
Elasmosaurus	III	10.3	3.7	3.29	0.0042

Hydrodynamic efficiency depends on the choice of a modern analogue. Bottlenose dolphins have an efficiency of 0·85 (Lang, 1975). Trout and salmon have an efficiency of 0·7-0·9 (Webb, 1975c). Ichthyosaurs may not have been as efficient as a dolphin, so assume an efficiency of 0·8, the middle of the trout and salmon range. Undulatory swimming is less efficient, estimated at 0·6 (Taylor, 1952), and this will be used for mosasaurs. Estimating the efficiency of modified underwater flying in plesiosaurs is more difficult. There are no data on sea lions, the obvious choice for a modern analogue. The efficiency of pectoral fin propulsion for the shiner perch, essentially an underwater flyer, ranges from 0·60 to 0·65 (Webb, 1975b). A value of 0·65 will be used for plesiosaurs. Substituting the appropriate values for the metabolic constant and hydrodynamic efficiency into eqn 7, and using the values of the drag coefficient and surface area of a prolate spheroid from Table 2, relative swimming speeds can be calculated. A summary of results is as follows: Bauplan I (ichthyosaurs): $U = 12\cdot2\ K$, Bauplan II (mosasaurs): $U = 9\cdot4\text{-}9\cdot6\ K$, Bauplan III (plesiosaurs): $U = 9\cdot8\text{-}10\cdot2\ K$, where K represents all the constants which are assumed to be equal for the three kinds of reptiles. Table 3 gives complete results and the contribution of each factor in eqn 7.

The relative values, not the numbers themselves, are important. For a given mass, ichthyosaurs were at least 25% faster than mosasaurs and 20% faster than plesiosaurs. Half the difference arises from differences in the estimates of the metabolic constant.

Table 3. *Relative maximum aerobic swimming speeds for selected marine reptiles. K is the product of all other factors in eqn 3, a constant for animals of the same mass.*

Genus	Speed U	Contribution of each factor		
		Metabolism $(a)^{0.3333}$	Efficiency $(e_h)^{0.3333}$	Shape $(1/(C_d\,A))^{0.3333}$
Ophthalmosaurus	12.2 K	3.07	0.93	4.27
Ichthyosaurus	12.2 K	3.07	0.93	4.28
Clidastes	9.4 K	2.73	0.87	4.09
Platecarpus	9.6 K	2.73	0.87	4.15
Peloneustes	10.2 K	2.73	0.87	4.31
Liopleurodon	10.2 K	2.73	0.87	4.31
Muraenosaurus	10.1 K	2.73	0.87	4.28
Elasmosaurus	9.8 K	2.73	0.87	4.16

Using a higher metabolic constant would increase the relative speed and increase the influence of metabolic rate factor in the calculation. Plesiosaurs were less than 5% faster than mosasaurs, the difference arising entirely from differences in shape (fineness ratio). Because of the shape factor, the long neck of the plesiosauroids resulted in a slower speed. Compare *Elasmosaurus* to *Muraenosaurus*: the difference in fineness ratio is due to the much longer neck in *Elasmosaurus*.

Substituting appropriate constants for K (assume a muscular efficiency, e_m, of 0·2; assume that wave drag doubles the drag coefficient) for animals of 2000 kg mass, approximate sustained swimming speeds can be calculated. The results suggest speeds of 4·8 m s^{-1} for the two ichthyosaurs, 4·0 m s^{-1} for *Liopleurodon*, *Peloneustes*, and *Muraenosaurus*, 3·8 m s^{-1} for *Elasmosaurus*, and 3·7 m s^{-1} for the two mosasaurs. These speeds are considerably higher than those calculated by Massare (1988).

This analysis is based on a number of assumptions (Massare, 1988), the two most important ones being, first, the drag estimate assumes that the animals can be approximated by a rigid body, and that movement of the tail or limbs does not contribute any additional drag. This is not the real situation. The drag coefficients taking propulsion into account can exceed predicted coefficients by a factor of 4–9 (Blake, 1983d; Lighthill, 1971; Alexander, 1977). This assumption is best for the plesiosaurs, which have a fairly rigid body. The effect of this assumption probably varies with the mode of propulsion and/or body shape. Second, the shape of the reptiles was assumed to be a prolate spheroid for calculating the surface area and in the expression for the drag coefficient. The assumption is better for mosasaurs, pliosauroids and ichthyosaurs than it is for plesiosauroids. Assuming another shape would obviously affect the results. The shape factor, however, accounts for only about a 5% difference in the speeds, so this assumption is probably less critical than the former.

CONCLUSION

The analysis presented here is admittedly a crude one, but it provides a quantitative estimate of relative swimming speeds and an indication of which factors contribute to the differences. For a given mass, ichthyosaurs were considerably faster swimmers than plesiosaurs or mosasaurs. This agrees with a more qualitative assessment based on gross morphological similarities with living dolphins. There is little difference in swimming speeds between mosasaurs and plesiosaurs of the same mass.

These differences, however, are based on animals of the same mass. A 2000-kg ichthyosaur with a fineness ratio of 4·0 would have been about 2 m long, whereas a 2000-kg mosasaur of fineness ratio 10·6 would have been about 3·8 m long (Figure 5). If differences are scaled to equal length, they are much more significant, and the difference between plesiosaurs and mosasaurs becomes important. Pliosauroids were faster sustained swimmers for a given length than the mosasaurs. Furthermore, the difference between long-necked plesiosauroids (e.g. *Elasmosaurus*) and shorter necked pliosauroids (e.g. *Liopleurodon*) for a given length is also significant, the latter being considerably faster. In terms of the prey that can be caught, the linear dimension of the predator (as relates to gullet width) may be more important than the mass.

The physiology of the animal is the most important factor in estimating sustained swimming speeds. Mosasaurs, related to living monitor lizards, were almost certainly ectotherms. More research on bone microstructure, however, is needed to provide better assessments of metabolism for the other reptiles which have no closely related

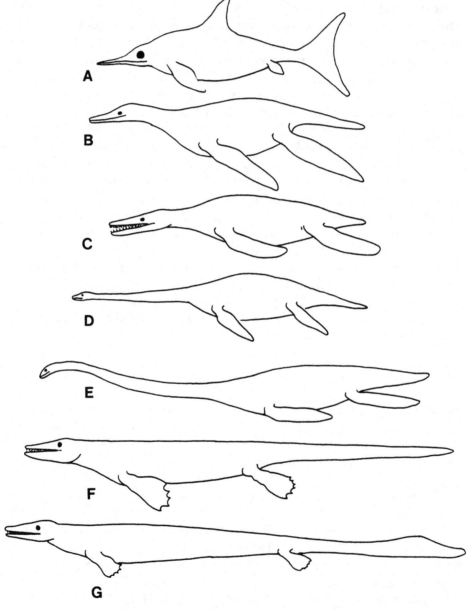

Figure 5. Reptiles used in swimming analysis, scaled to the same mass (2000 kg). Mass calculation approximates the shape as a prolate spheroid. (A) 1·9-m long ichthyosaur, *Ophthalmosaurus*; (B) 2·3-m long pliosauroid, *Peloneustes*; (C) 2·2-m long pliosauroid, *Liopleurodon*; (D) 2·8-m long plesiosauroid, *Muraenosaurus*; (E) 3·7-m long plesiosauroid, *Elasmosaurus*; (F) 3·8-m long mosasaur, *Platecarpus*; (G) 4·2-m long mosasaur, *Clidastes*.

living relative. Did the plesiosaurs have an ectothermic metabolism or was it more like that inferred for the ichthyosaurs? Was there a difference between plesiosauroids and pliosauroids? Answers to these questions may lie in the microstructure of fossil bones and could significantly alter the results of the analysis presented here.

This analysis strengthens the hypothesis that mosasaurs (and other long-bodied marine reptiles) and post-Triassic ichthyosaurs were different kinds of predators (Massare, 1988). Calculations suggest that a 2-m long ichthyosaur swam 25% faster than a 3·8-m mosasaur. The former would have had a significant advantage in long-distance hunting and pursuit, when sustained swimming is important. The burst speed, however, is proportional to body length, measured at a maximum of ten times the body length for fish (Bainbridge, 1960). Thus the burst speed of the mosasaur would have been almost twice that of the ichthyosaur, giving the former a significant competitive advantage as an ambush predator that relied on quick lunges to capture prey.

The Society for Experimental Biology provided funds which enabled me to attend the symposium. I thank Dr M.A. Taylor for his thoughtful review and for providing additional references. Mr A.C. Benton translated the Klima papers and they were kindly made available to me by Dr M.A. Taylor.

Chapter 11

Stone, bone or blubber? Buoyancy control strategies in aquatic tetrapods

MICHAEL A. TAYLOR

National Museums of Scotland, Chambers Street, Edinburgh,
EH1 1JF, Scotland

An air-breathing tetrapod's buoyancy depends in part on the animal's contained gas, and therefore on the ambient pressure and depth. An animal can vary gas content as a hydrostatic control mechanism to reach neutral buoyancy at a particular depth. Alternatively, or additionally, it can compensate for buoyancy hydrodynamically, for example by producing lift forces during swimming: this is most effective for animals which contain less gas, and thus swim faster and deeper. Two extreme evolutionary adaptations can be predicted from this, and have indeed appeared through convergent evolution: body density and submerged lung-volume reduction for fast, cruising swimmers and deep divers (cetaceans, phocid pinnipeds and ichthyosaurs) and body density and submerged lung-volume increase for bottom-walkers (sirenians and placodonts). A third strategy, swallowing stones, maximizes the flexibility of hydrostatic control and enables the identification of a third case of convergent evolution in the manoeuvrable underwater flyers (otariid pinnipeds, penguins, and plesiosaurs).

INTRODUCTION

The marine reptiles, birds and mammals include many of the dominant marine predators and filter feeders of the last 250-odd million years (Carroll, 1988; Massare, 1988). Control of buoyancy is an important adaptation to life in water, yet little is known about how tetrapods in general do so. Although there are important studies on particular groups (some cited in this chapter), there is no one synthesis which explains the bewildering, indeed, apparently contradictory, variety of adaptations commonly ascribed to the need to control buoyancy. Why should manatees and dugongs have

Maddock, L., Bone, Q. & Rayner, J.M.V. (ed.). *Mechanics and Physiology of Animal Swimming.*
© 1994. Cambridge University Press.

massively heavy skeletons, while the skeletons of modern whales are in fact much lightened, and otariid sealions and penguins swallow pebbles? And why should their respective Mesozoic analogues, the placodonts, ichthyosaurs, and plesiosaurs, show exactly the same apparent strategies of buoyancy control? In this chapter, I aim to present such a synthesis and to show that these three strategies can be predicted from a knowledge of the basic physics involved, and that they are ecologically the most appropriate for each group.

This chapter is a brief treatment and I hope to present a fuller analysis elsewhere, while I have already discussed gastroliths (stomach stones) in detail (Taylor, 1993). Alexander (1983, 1990) outlines the physics involved, while physical data are taken from Wainwright *et al.* (1976) and Weast *et al.* (1988).

BUOYANCY CONTROL IN AN AIRBREATHING TETRAPOD

From Archimedes' Principle, an animal in water is buoyed up by an upthrust, its buoyancy, equal and opposite to the weight of water which it displaces. The difference between this upwards buoyancy and the animal's weight in air is the resultant vertical force on an animal, its net weight in water or its net buoyancy (henceforth just 'buoyancy'). If the animal is exactly the same (average) density as water, then its own weight exactly balances the buoyancy and it is neutrally buoyant. But if it is denser, it is negatively buoyant and tends to sink; if it is lighter, it is positively buoyant and tends to rise. If an animal wishes to remain at a particular depth, it has two options. It can control its depth hydrostatically, adjusting its own density to achieve neutral buoyancy, or it can control its depth hydrodynamically, using vertical forces generated by swimming movements, such as lift produced by a horizontal hydrofoil fin during forwards swimming. Of course, the animal can coarsely adjust its buoyancy and then use hydrodynamic forces to compensate for any residual excess positive or negative buoyancy.

Most of the components of a tetrapod's body are denser than water (Table 1), and only air and lipid (including fats and waxes) are lighter than water. However, each has functions other than buoyancy control: the skeleton has a mechanical role (Currey,

Table 1. *Densities and buoyancies of various substances.*

Material	Density / 10^3 (kg m^{-3})	Buoyancy in water / 10^4 N m^{-3}
Air	0.0	1.0
Freshwater	1.00	0.00
Sea-water	1.02	
Lipids (fats, oils and waxes)	0.85-0.95	0.15-0.05
Muscle	1.05	-0.05
Cartilage	1.1	-0.1
Bone (solid)	≤2.0	-0.6-1.0
Basalt	2.4-3.1	-1.4-2.1
Granite	2.7	-1.7
Quartz	2.65	-1.65

1984), lipid is used for energy storage, thermal insulation and streamlining (Pond, 1978), and the primary role of air is in respiration, so we should not expect different animals to have the same balance of tissue composition. Indeed, as I describe below, different buoyancy control strategies demand different body compositions.

The actual buoyancy of an animal depends on the mean density of the animal, and therefore on the precise volumes of the different constituents. Amongst the most important, because of its negligible density and great positive buoyancy, is the air in the lungs, yet this factor is also the most difficult to measure. First, because of the inherent problems of measuring it in life rather than after death, and more specifically during a dive, and, second, because gas volume changes rapidly as pressure increases with depth. Neglect of either factor will almost completely negate any analysis of buoyancy.

Total buoyancy depends on the balance between positively and negatively buoyant components. It is highly sensitive to small changes in these, caused most obviously by the compression of air during a dive, but also by factors such as overall body composition, especially the density and amount of bone, and the amount of lipid tissue, which itself often varies seasonally, for example in a migrating whale. Relatively minor factors such as the variation of sea-water salinity, changes of density of solids and liquids with pressure and temperature, and the departure of gases from ideal behaviour at high pressures, must be taken into account when considering an individual animal's buoyancy (e.g. Clarke, 1978a,b,c). However, such changes are readily compensated for by minor changes in the amount of air retained in the lung, and I do not discuss them further. This chapter is, rather, concerned with the gross variation in body composition and buoyancy control in different taxa. I further simplify it by assuming that the density of solids and liquids do not change significantly, that any gas behaves as an ideal gas, and that temperature is constant.

The change of pressure with depth affects the volume of any gas in the respiratory system, as well as any in the gut and in the integument (insulation amongst the fur or feathers). At a depth, D, the underwater pressure, P, is equal to the atmospheric pressure at the surface, A, added to the weight per unit area of the overlying water, which is itself equal to $g\rho_w D$ (i.e. the weight of a unit volume of water times the height of the water column, where g is the acceleration due to gravity, and ρ_w is the density of water). It can be shown that the net buoyancy of an animal, F_T, depends on the constant buoyancy of its solid and liquid constituents, F_S, plus a factor for the buoyancy of its gaseous content. If v is the amount of gas measured in moles (absolute quantity), R is the gas constant, and T the absolute temperature, then:

$$F_T = F_S + v R T g \rho_w / (A + g \rho_w D). \tag{1}$$

This equation shows how an animal can become neutrally buoyant ($F_T=0$) by compensating negative buoyancy, F_S, from its solid and liquid components, with positive buoyancy from its gas content. All other things being equal, neutral buoyancy can be achieved by a high gas content in an otherwise relatively dense animal, or a low gas content in a not very dense animal. In particular, the buoyancy due to gas (the second factor on the right-hand side) is directly proportional to v, the amount of gas, as R, T, A, g and ρ_w are all constant. It is also dependent on depth (to be more specific, it is a hyperbolic function of depth, Figure 1). In practical terms, the buoyancy due to gas is greatest at the surface and decreases very rapidly at first, but then more slowly with increasing depth. Each added 10 m of depth adds 100 kPa, almost exactly one atmosphere. Therefore the first 10 m depth doubles the pressure and halves the gas volume and thus the buoyancy due to gas. In other words, however deep an animal dives, at least half the total change of buoyancy will take place within the first 10 m depth. Further increases in depth have increasingly small effects; for example, diving from 100 to 110 m will decrease buoyancy by about 9%.

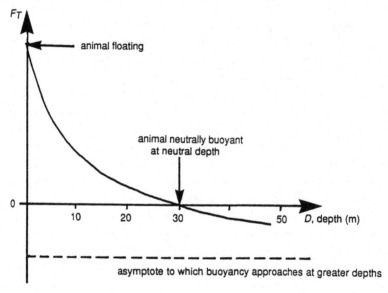

Figure 1. One possible relationship of buoyancy, F_T, with depth, D, of an air-breathing tetrapod, which is denser than water, without any air in its lungs, but which floats at the surface when the lungs are inflated. See text for discussion. Neutral depth just happens to be 30 m under the circumstances of this particular example.

If the animal is not too dense, it floats at the surface. But when it dives, its buoyancy decreases, and if it dives deep enough, its buoyancy becomes neutral and then increasingly negative. I here term the depth where it is neutrally buoyant the 'neutral depth'. The animal is then neutrally buoyant and can hover stationary in mid-water. However, this is an unstable equilibrium, as any perturbation above or below the neutral depth will cause a change in buoyancy, tending to increase the perturbation. If an animal is stationary at a depth shallower than the neutral depth, it rises to the surface under buoyancy. If it is stationary below the neutral depth, it sinks. In both cases, the further it gets from neutral depth, the stronger the forces become.

An animal can adjust its neutral depth by changing the amount of gas retained in its lungs (Figure 2), or by ingesting dense objects such as stones. More air or a lighter body gives a deeper neutral depth, and less air or a denser body gives a shallower neutral depth. Thus any diving animal can adjust its buoyancy by adjusting the amount of air retained within its lungs so that it can float at the surface, hover at a fixed neutral depth, or reach a third equilibrium on the bottom of the water body, if this is below neutral depth: the animal can then rest on the bottom with an appreciable reaction force, providing useful friction against water currents or the forces involved in removing prey from the substrate. However, changing the amount of gas is not equivalent to changing the buoyancy of the solid and liquid components. An animal which is still positively buoyant at a particular depth can attain neutral buoyancy by swallowing, say, some stones, in which case the amount of gas and the rate of loss of buoyancy with increasing depth remain the same, or it can exhale some gas in which case the rate of loss of buoyancy with increasing depth is decreased in proportion. These lead to two different strategies, as I discuss below. (In graphical terms, changing the amount of gas will change the shape of the curve, as in Figure 2, but changing the buoyancy of the animal's solid or liquid constituents will move the curve up or down on the graph's axes.)

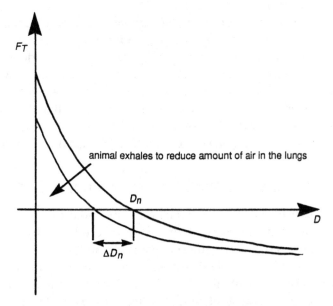

Figure 2. As Figure 1, but showing buoyancy against depth for the same animal on two separate dives with different amounts of air in the lungs. D_n, neutral depth, is reduced when the amount of air is reduced.

The optimum neutral depth will depend on the circumstances. The optimum is obvious if the animal is feeding at a given depth, swimming slowly, or cruising deep enough to be outside the effects of surface turbulence and the production of surface waves. Alternatively, if the animal needs to rest or walk on the bottom, or uses friction to counteract water currents, neutral depth should be set at some point above the bottom of the water column.

Here are some simple examples of adaptive buoyancy control; I give some more later. The darter, *Anhinga melanogaster*, becomes approximately neutrally buoyant and can remain with little or no movement at 1 to 4 m depth, a very useful adaptation as it stalks its prey in mid-water (Hustler, 1992). The pelagic yellow-bellied sea snake, *Pelamis platurus* (Graham *et al.*, 1987a), sets the appropriate volume of gas in its lungs before diving to achieve nearly neutral buoyancy at the depth at which it desires to swim. *Pelamis* also spends much time floating, drifting on surface currents towards prey-rich surface convergences, and its surface buoyancy provides it with an anchor for holding its body as a fulcrum for striking at prey (Dunson, 1975; Graham *et al.*, 1975; Kropach, 1975; McCosker, 1975). In contrast, many other species of sea snakes crawl, rest or swim slowly over the bottom on, for example, coral reefs (Heatwole *et al.*, 1978), and presumably control their buoyancy (Heatwole & Seymour, 1975). This slow swimming enables them, for example, to search for fishes hiding in crevices (McCosker, 1975). The freshwater chelonian, *Pseudemys scripta elegans*, can adjust its buoyancy very precisely, despite the experimental addition of lead weights or plastic floats, by reciprocally changing the lung volume and the amount of water stored in cloacal bursae within the rigid shell (Jackson, 1969).

Juvenile animals might be expected to be especially positively buoyant, as a safety measure to prevent drowning while their swimming abilities are not yet fully developed. Calves of the bottlenose dolphin, *Tursiops truncatus*, are indeed born with an especially high proportion of blubber, and observations of one calf showed that it was unable to reach neutral buoyancy until about six months of age (Cockcroft & Ross, 1990). Hatchling green turtles, *Chelonia mydas*, are so buoyant that they cannot dive or sustain underwater swimming. They are thought to have a neustonic lifestyle, carried by surface currents and feeding on surface-dwelling jellyfish (Davenport *et al.*, 1984).

HYDROSTATIC VS HYDRODYNAMIC BUOYANCY CONTROL

To swim at a particular depth, an animal can either adjust its buoyancy so that it reaches neutral buoyancy at this point, or else it can exert a hydrodynamic force to compensate for any positive or negative buoyancy. It can direct this force upwards or downwards to remain in one place. Alternatively, if it is swimming horizontally forwards, it can adjust its line of thrust to include a vertical component of force. This is energetically expensive because of induced drag. However, this may be offset by avoiding the drag penalty associated with the increased volume of air, or a potentially even greater volume of solids or liquids needed to correct buoyancy hydrostatically. It may also be useful in minimizing the amount of contained air, and therefore the change of buoyancy with depth.

Suppose that an animal can swim most economically if it controls its buoyancy by exerting an upwards or downwards force up to or equal to some value, H. The animal then no longer has one neutral depth but a range, which I call the 'neutral band' (Figure 3). The effective size, y, of this band depends on how quickly buoyancy is changing with depth. At the surface, the curve is steepest and the rate of change greatest, so that y is smallest for a given H. Also, for greater amount of gas v, y is smaller. So, for a given H, y is greatest for greater depths and smaller amounts of gas. This suggests that animals are most likely to use hydrodynamic buoyancy control when they are diving deeply or when they contain little gas, or both.

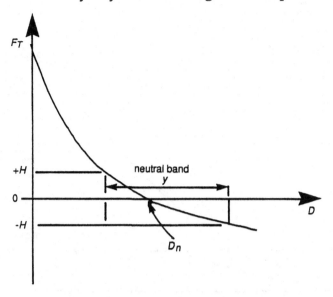

Figure 3. The effect of hydrodynamic buoyancy control in a case where the animal can exert a vertical force of magnitude up to or equal to H to correct excess upwards or downwards buoyancy. The effect is to increase the range of depths, or *neutral band*, over which the animal can swim horizontally.

When is hydrodynamic compensation superior to hydrostatic control? Alexander (1969, 1972, 1977, 1990) and Gee (1983) discuss the energetic penalties (in terms of added drag at different speeds) of buoyancy control adaptations in fishes. I summarize and adapt their conclusions here for the case of an animal which has a certain upwards buoyancy to correct (Figure 4). Hydrodynamic buoyancy control gives a fixed drag penalty whatever the speed, assuming a constant ratio of lift to drag. However, adding air, bone or stones in a hydrostatic control mechanism increases the volume, and therefore the surface area and cross-sectional area, of an animal, and hence the drag. As the drag on an animal is roughly proportional to the square of the speed, hydrostatic control induces an increasing penalty with increasing speed. Furthermore, the use of an inefficient ballast, such as bone, rather than a denser ballast such as stone (see below), requires the addition of extra volume for a given change in buoyancy. This leads to the prediction that stone, and to a lesser extent bone, ballast are more efficient at lower speeds, and hydrodynamic control is more efficient at higher speeds. I do not attempt a quantitative analysis here; however, it is interesting that the slowest marine mammals, the sirenians, do indeed use bone, the much more mobile otariid pinnipeds (and penguins) use stones, and the fastest of all, the cetaceans, appear to use neither (Taylor, 1993).

WORK DONE DURING A DIVE

Buoyancy is a gravitational phenomenon and the work done against it produces a form of gravitational potential energy. If an animal dives to a maximum depth, deeper

M.A. Taylor

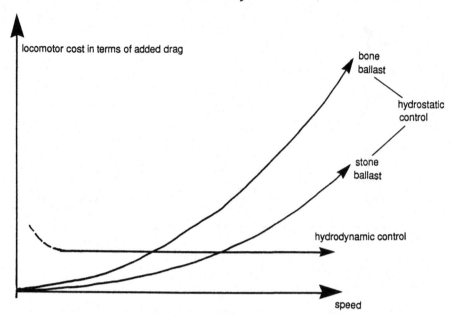

Figure 4. Diagram showing the relative merits, in terms of added drag, of different buoyancy control strategies to correct a given excess positive buoyancy. See text for discussion.

than the neutral depth, and back again, then work against buoyancy is done diving as far as the neutral depth. Below the neutral depth, the animal is negatively buoyant and sinks. However, when the animal swims upwards again, it has to do work to restore the potential energy consumed by diving below neutral depth. Above the neutral depth the situation is reversed and the animal tends to rise, expending the potential energy created during the dive from the surface to the neutral depth. The actual metabolic work done by the animal will depend on the efficiency of conversion of metabolic energy to locomotor thrust. Furthermore, the actual work also depends on the speed of the dive, and must take into account locomotor drag and any changes in momentum and kinetic energy. A fuller analysis, and therefore optimization, would also include the relative costs of horizontal locomotion at different depths. See Lovvorn & Jones (1991a,b), Hustler (1992) and Wilson *et al.* (1992) for case studies of diving birds which consider these and related problems. Also, in those birds and mammals where air trapped in the integument is used to help provide heat insulation, a further conflict arises between energy wasted due to excess buoyancy from this air vs energy lost through inadequate insulation (Wilson *et al.*, 1992).

So far I have assumed that animals dive vertically and then swim horizontally. However, actively swimming animals can minimize the total cost of locomotion by varying their depth as they swim along, exploiting the interchange of kinetic and potential energy, and the differences in drag between swimming and gliding. One case is the porpoising of some penguins, some otariids and some odontocetes: they progress by alternately swimming powerfully forward, leaping out of the water, and diving back in, exploiting the relatively low drag of a moving object in air (Blake, 1983c; Blake &

Smith, 1988). Another case is the strategy of some fish which are denser than water. They swim actively forwards and slightly upwards, then glide downwards under gravity and momentum, the next burst of swimming returning the animal to the initial depth (Weihs, 1973). An inversion of this strategy, swimming downwards and gliding upwards, might work just as well for a positively buoyant tetrapod, especially if the upwards/downwards cycle were coupled with the pre-existing need for periodic breaths at the surface.

Thus, if an animal is swimming fast enough for forwards motion to be a significant component of its expenditure in locomotor energy, then we must expect complex interactions between buoyancy and locomotor efficiency, where there is no one optimum swimming depth. Thus the simple strategy of setting (by appropriate exhalation) and then remaining at a single fixed neutral depth may be valid only for stationary, or slowly swimming, animals. Each case must be studied on its own merits.

EVOLUTIONARY STRATEGIES OF BUOYANCY CONTROL

The buoyancy of an animal can be changed by altering the volume and density of its non-gaseous constituents. However, these are radical changes which involve many other functions such as locomotion and energy storage, and can take place only on an evolutionary time scale, with the exception of swallowing and vomiting stones (and not entirely so even then; Taylor, 1993). Also, they imply commensurate changes in gas content, or the animal will become hopelessly light or dense. There are three strategies open during the evolution of an aquatic tetrapod: (1) to reduce body density and air quantity, (2) to increase body density and air quantity by accumulating extra bone, and (3) to increase density and air quantity by swallowing stones.

Minimizing density and air quantity

Reducing body density, by reducing skeletal mass and perhaps increasing the proportion of lipid, allows a reduction in gas content. It reduces the work needed to dive to a given depth. The loss of gas volume allowed by mass reduction would be useful to a deep-diving animal (see also Walls, 1983). The decrease in the absolute range of buoyancy forces, and in their rate of change, increases the depth of the neutral band: in other words, the animal can swim over a wider range of depths for the same maximum vertical force exerted, maximizing the efficacy of hydrodynamic buoyancy control as noted earlier. This would be most useful to a cruising pelagic animal, such as an open-sea predator, filter feeder or migrator, which habitually swims at a significant speed, and therefore uses hydrodynamic rather than hydrostatic control.

There is indeed evidence for reduction of bone density and overall mass in at least some cetaceans, which are pelagic cruising forms, and some ichthyosaurs, which are thought to have had a similar lifestyle (de Buffrénil et al., 1985, 1986; de Buffrénil & Mazin, 1990; Walls, 1983). De Buffrénil and co-workers interpret this as an adaptation to reduce inertia and improve acceleration (cf. Webb & Skadsen, 1979). I suggest the alternative (and not necessarily exclusive) hypothesis of buoyancy control, as does

McGowan (1992; see this and Taylor 1987 for a discussion of buoyancy control mechanisms in ichthyosaurs).

Maximizing density and air quantity

Increasing body density by increasing skeletal mass requires a compensating increase in lung volume. As a result, buoyancy is lost much more rapidly with depth. In an extreme case, the animal can lose buoyancy so quickly that, although floating when at the surface, it becomes negatively buoyant at the bottom in even very shallow water. This would be useful to a slow-swimming animal bottom-walking while feeding on stationary benthic food. The large lung volume is potentially available for use as an oxygen store, which itself implies shallow diving because of the inability of the lung to exchange gases when collapsed at depth, and possibly also the risk of decompression syndrome. The increase in work done during a dive, and the increase in rate of change of buoyancy with depth, penalize shallow divers and slow swimmers far less than deep divers or fast swimmers. Indeed, they can be useful to benthic feeders.

The living sirenians fit this model well. They show a condition commonly referred to as pachyostosis, the development of thicker bones, with reduced or absent marrow cavities, less porosity, and sometimes some increased mineral content (see e.g. Kaiser, 1960; de Buffrénil & Schoevaert, 1989; de Buffrénil & Mazin, 1989; Domning & de Buffrénil, 1991, and note the terminological discussion in the latter paper). They have enlarged lungs which remain fully inflated during the dive (Hartman, 1979; Caldwell & Caldwell, 1985; Nishiwaki & Marsh, 1985). They can float, hover in mid-water at neutral depth just below the surface, and rest and walk while negatively buoyant on the bottom, where they graze on sea-grasses in very shallow water (3-4 m).

The marine sea otter, *Enhydra lutris*, has apparently adopted the same strategy of buoyancy control. It has a benthic invertebrate diet and crushing dentition, it often feeds and rests while floating at the surface, and it has, at least in parts, a relatively dense skeleton (Fish & Stein, 1991; however, this can be supplemented by carrying an anvil stone, functionally analogous to gastroliths). Furthermore, its exceptionally large lungs both provide two-thirds of its oxygen reserves (Lenfant *et al.*, 1970) and cause it to lose buoyancy with depth rapidly. Finally, it holds even more air, trapped in its fur for thermoregulation, rather than using lipid for insulation, which is further evidence that *Enhydra* adopts this strategy of buoyancy control. *Enhydra* appears to be a modern ecological analogue of the placodonts, a group of amphibious Triassic marine reptiles which had massive crushing teeth and jaws and fed on hard-shelled benthic invertebrates in shallow water (Carroll, 1988; Pinna & Nosotti, 1989). It is not clear whether the placodonts often developed pachyostosis, but some carried additional bony armour, and at least one genus, *Placodus*, had additional development of bone on the transverse processes (Kaiser, 1960).

Gastroliths (stomach stones) are denser and more efficient ballast than bone (Taylor, 1993). The density of quartzite, for example, is about 2650 kg m^{-3}, compared to bone at about 2000 kg m^{-3} (Currey, 1984), so that a given volume of stone gives as much negative buoyancy as one and a half times the volume of bone, with economies of body mass and volume and therefore swimming efficiency. If stones are available, they avoid the metabolic cost of depositing bone, and can be eaten or vomited to change buoyancy quickly, unlike bone.

It has long been disputed whether stomach stones or 'gastroliths' in aquatic tetrapods were used for processing food or buoyancy control (e.g. Cott, 1961; Davenport et al., 1990). Taylor (1993) reviewed the evidence and noted that their occurrence shows no correlation with diet, once allowance is made for accidental ingestion by benthic feeders; instead, they are usually found only in crocodilians, which use them to help lurk in ambush under the surface of the water (Cott, 1961), or in the penguins, the otariid pinnipeds, and the plesiosaurs, which are all predatory underwater 'flyers' (Alexander, 1989a; Clark & Bemis, 1979; Feldkamp, 1987; Riess & Frey, 1991). Gastroliths do not usually occur in tetrapods using fish-like caudal fins (cetaceans, ichthyosaurs, mosasaurs), or hind limbs modified to act as caudal fins (phocid pinnipeds, Fish et al., 1988, and the walrus, *Odobenus*, Gordon, 1981). This distribution suggests a locomotor function, presumably in buoyancy control.

I suspect that the plesiosaurs, penguins and otariids exploit the flexibility and efficiency of gastroliths to provide hydrostatic control efficient at low speeds, where the manoeuvrability of these animals (Godfrey, 1985; Hui, 1985) due to their long necks and vectored thrust can best be used. I do not know if these animals are particularly dense or particularly light, or fall between the pachyostotic forms and the lighter cetacean-ichthyosaur strategy. If so, they form a third buoyancy control strategy. Further research is needed here, and also to determine the relative merits of underwater flight versus caudal fin propulsion.

CONCLUSION

Verifying the hypotheses which I have set out will need careful studies on a wide range of aquatic tetrapods. However, they may provide a useful contribution to refining locomotor and energetic strategies in living forms, and go some way towards explaining striking features such as pachyostosis and gastrolith swallowing in both living and extinct forms. It is encouraging that such a simple model coherently ties together recent case studies on buoyancy control on individual taxa. Indeed, it has provided new evidence for functional mechanisms underlying the classic example of convergent evolution between ichthyosaurs and cetaceans, as well as two further cases of convergence: the shallow benthic feeding sirenians, sea otter and placodonts, and the gastrolith-using underwater flyers, the plesiosaurs, otariids and penguins.

This research was supported by a Leverhulme Trust Research Fellowship and by Leicestershire Museums, Arts and Records Service. I am grateful to the University of Leicester for facilities, and to Frank Fish, Judy Massare and Rory Wilson for discussion.

Chapter 12

Functional anatomy of the 'flight' apparatus in penguins

R. BANNASCH

Bionik und Evolutionstechnik, Technische Universität, Berlin,
Ackerstraße 71-76, D-1000 Berlin 65, Germany

Penguins are the best-adapted birds to wing-propelled diving and swimming. In water the weight of their perfectly streamlined body is balanced by buoyancy. Hence the strong 'flight' apparatus is used only for thrust production. Similarly to flying birds penguins flap their wings. But, contrary to fast flight in air, thrust is produced under water during the up-stroke as well as during the down-stroke. Corresponding wing profiles are developed. Since the wings are active in a permanently extended position, the shoulder joint is simplified. The examination of its structure and kinematics revealed a new type of articulation. The 'rotation' of the head of the *humerus* in the shoulder joint can be described as a two-axis-eccentric motion conducted by the *ligamentum acrocoracohumerale* and an articulation surface of the *caput humeri* shaped like a logarithmic spiral. This model led to new interpretations of the function of the muscle system. Besides the preponderant thrust-producing (flapping) muscles, working mainly isotonically, muscles can be distinguished which are managing the transfer to the body of the forces produced, thereby operating isometrically. Another group of muscles must control the position of the *humerus*, adjusting the hydrodynamic angle of attack corresponding to respective flow conditions. In kinematic studies conducted on live penguins, different modes of wing action were observed. The functional anatomical demands of the balance of forces and momentum are discussed with respect to the non-steady flow around the wings.

INTRODUCTION

A diverse range of animals use a special mode of swimming which, because of some similarities with the flight in air, is called 'aquatic flight'. In this group it is of particular interest to analyse the locomotor system of wing-propelled diving birds which were first (and foremost) adapted to flight and in which, according to Storer (1960), "modifi-

Maddock, L., Bone, Q. & Rayner, J.M.V. (ed.). *Mechanics and Physiology of Animal Swimming.*
© 1994. Cambridge University Press.

cations for an aquatic existence have been superimposed on this plan with varying success". So the question is, to what extend should, and can, the flight apparatus of birds be modified to meet the demands of the denser medium of water?

Of the wing-propelled divers, the penguins are by far the most specialized. The special structure of their wings, the peculiarities of the plumage, the torpedo-like shape and solid construction (high density) of the body etc. suggest that their complete change from aerial to underwater flight has been realized without any compromises.

To understand the peculiarities of the functional anatomy of the flight apparatus in penguins, it is helpful to have a brief review of the physical constraints faced by wing-propelled diving birds.

PHYSICAL CONSTRAINTS AND THE PRINCIPLE
OF THRUST GENERATION

The changeover from aerial to aquatic flight poses complex physiological and physical problems. The locomotor system of flying birds is already one of the most efficient in animals in respect of its energy turnover, and the physiological mechanisms may not be able to be further improved in favour of an increased diving capacity (for details see p.185). Thus, for example, the oxygen storage capacity can be increased extensively only by adding mass and volume (size progression). At the same time, the density of the body must be adapted to that of water by changing from an 'ultra-light' to a more solid construction. On the other hand, even if the wings are partly folded under water, the denser medium requires a reduction of the wing area, or at least prevents their enlargement proportionally to the weight progression. As the volume (or mass) varies with the cube, and the wing area with the square, of linear proportions, a geometric enlargement of the bird leads to an increased wing loading. To keep flying in air, the larger bird must either fly faster than the smaller one of similar shape (Storer, 1960), which decreases its manoeuvrability, or its wings must be enlarged allometrically. In fact, from the discussion of Dinnendahl & Kramer (1957) and Meunier (1958) it became clear that within the recent groups of birds (e.g. gulls) the tendency is rather the opposite: with increasing size of the bird the wing area becomes proportionally reduced. Respective data on Alcidae were published by Livezey (1988). Continued size progression would finally lead to the loss of flight capability if an abrupt change of the geometric proportions, called 'transposition' by Meunier, is prevented, for example by the circumstances cited above, or if it simply does not occur. In principle the loss of flight may occur at every stage of the size progression of a given type of construction.

I agree with Simpson (1946) and Storer (1960) that it is most likely that the early penguins, deriving from an ancestral morphotype similar to that of the recent diving-petrels (Pelecanoididae), may have passed through a functional (not phylogenetic!) stage quite similar to the guillemots or murres (*Uria*) and the razorbill (*Alca*), which are the largest recent birds able to manage both flight and subaquatic wing propulsion.

The broad and partly controversial discussion in the literature on the origin and taxonomic relationship of penguins, which was based not only on the unique wing morphology but included the many other peculiarities of the skeleton-muscle system, as well as feather structures, biochemical and also behavioural components, was reviewed by Bannasch (1986a,b, 1987). Most studies confirm a close relationship with the Procellariiformes, but some (mainly biochemical) similarities with the Gaviiformes, Podicipediformes, Pelecaniformes and Apterygiformes were found as well. At last, based on data obtained by the DNA-DNA hybridization technique, Sibley (1991) rearranged the frigatebirds, penguins, loons, petrels and albatrosses in a single superfamily Procellarioidea.

Considering the biophysical aspects of evolution, Stonehouse (1975) concluded that the ancestors of the penguins, which first ceased flying in air, could not have been larger than the smallest living penguins (*Eudyptula minor*).

Once exclusively in the water, the problem of the surface/volume ratio does not persist. In penguins the body weight is slightly overcompensated by the hydrostatic buoyancy (Mordvinov, 1980), which is somewhat variable with depth (Wilson *et al.*, 1992). Buoyancy is not a significant factor, and hence the strong flight apparatus is used only for thrust production to overcome the fluid resistance, *D*, of the body, which may be written:

$$D = q\,C_D\,A \tag{1}$$

where *D* denotes the drag force (in Newtons) experienced by the body, $q=(\rho/2)v^2$ is the dynamic pressure (in N m^{-2}), where ρ denotes the density of the water (kg m^{-3}) and *v* the velocity (m s^{-1}), C_D denotes the drag coefficient (dimensionless), and *A* denotes the reference area of the body (m^2). The drag coefficient may vary moderately with the Reynolds numbers Re, given by

$$\mathrm{Re} = (v\,l) / v \tag{2}$$

where *v* denotes the free stream velocity (m s^{-1}), *l* the reference length of the body (*m*), and *v* the kinematic viscosity of the water (m^2 s^{-1}).

This variation is discussed below, but for the moment I shall take it as constant to illustrate the effect of size progression on the kinematics of the aquatic flight. With this simplification, the body drag as well as the wing area are square functions of linear dimensions. Thus the birds are not limited in maximum size by the surface/volume ratio.

Penguins flap their wings in a similar way to flying birds. But contrary to fast flight in air, thrust is also produced during the up-stroke. The penguin wing certainly acts as a hydrofoil, but the physical principle of force production (profile action) is similar to that of an aerofoil in flight (cf. Rayner, 1985). For this reason I feel free to use the term 'wing' for the forelimb of the penguin as well.

Suppose three model-penguins are geometrically similar but of different size. The smallest (S) has a body length like that of a little penguin (~30-35 cm), the medium sized (M) may represent a gentoo (~65-70 cm) and the largest (L) has a length of ~95-100 cm and should represent an emperor penguin. Thus the respective length ratios are approximately 1:2:3.

Suppose the three penguins swim at the same speed. Then the ratios of the body drag are $D_S:D_M:D_L=1:4:9$. Consequently, the larger the penguin, the more thrust has to be produced by the wings. If the wings are geometrically similar (ratios of wing areas 1:4:9 respectively), and if the principle of force generation does not change, so the respective thrust forces can be produced if a similar advance ratio λ is achieved. That is

$$\lambda_S = \lambda_M = \lambda_L = \frac{u}{\omega_S R_S} = \frac{u}{\omega_M R_M} = \frac{u}{\omega_L R_L} \tag{3}$$

where u denotes the swimming velocity, ω the angular velocity of the wing, and R the radius of the wing (length from the *caput humeri* to the wing tip).

If ratios of $R_S:R_M:R_L=1:2:3$ are given for the wing length, so the ratios of the angular velocities must be the opposite ($\omega_S:\omega_M:\omega_L=3:2:1$), as must the respective ratios of the wing-beat frequencies. In other words, to swim at the same velocity, a smaller penguin has to flap its wings more often than a larger one: with size progression the wing-beat frequency decreases. If the maximum work (force x distance) performed per unit volume of muscle in one contraction is independent of the size of the animal (Schmidt-Nielsen, 1977), the power output during contraction will be a direct function of the speed of shortening, or strain rate. Thus power output is scaled to decrease with increasing body size.

This consideration is theoretical, but it explains both the general tendency of the data on swimming kinematics of penguins published by Clark & Bemis (1979), and the vast differences in the diving capacity of penguins of various sizes. In nature, however, considerable deviations from the model may result from variation of the wing-beat amplitude (stroke angles), different modes of wing action and interposed gliding phases etc., and also from morphological (Livezey, 1989) and biochemical (Baldwin, 1988) differences between individuals and species.

Apart from thrust generation, the flapping wings produce positive and negative vertical forces during the down-stroke and up-stroke, respectively. In steady submerged swimming, the impulses perpendicular to the direction of translational motion must be balanced. Therefore, the simplest solution would be mirror-symmetry of the up- and down-strokes, as has been described for *Eudyptula minor* by Clark & Bemis (1979). However, the few kinematic pictures available from underwater flight of other penguins (Clark & Bemis, 1979; Hui, 1983), and the results of my own observations on *Pygoscelis adeliae*, *P. papua*, and *P. antarctica* (unpublished) show the down-stroke to have a greater potential for thrust production than the up-stroke. Usually the up-stroke lasts longer. Then, in most cases, the stroke plane is somewhat tilted forwards. Is this asymmetry a relic from the flying ancestors? The recently available data on swimming kinematics do not allow general conclusions.

At this stage, a clarification of the general morphological and anatomical structures of penguins for underwater flight is important.

MATERIAL

The anatomy of the active and passive apparatus of movement was studied by dissection of 24 individuals of *Pygoscelis papua* (9), *P. adeliae* (9) and *P. antarctica* (6) collected at King George Island, Antarctica. Some comparative data (mainly on muscle mass) were obtained from *Eudyptes chrysolophus* (1) and *Aptenodytes forsteri* (1) collected in the same region.

In the anatomical terminology I follow Baumel *et al.* (1979). In some cases, when the dictionary was incomplete, I refer directly to Fürbringer (1888).

EXTERNAL OBSERVATIONS AND HYDRODYNAMIC CHARACTERISTICS

Trunk

In the swimming posture, the body shapes of the pygoscelid penguins resemble one another in being spindles with high values of maximum thickness position (44-47%) and thickness ratio ($0.22 < d/l < 0.25$) (for details see Oehme & Bannasch, 1989; Bannasch, in press). A small dorso-ventral asymmetry is evident from the lateral view (Figure 1A). It is likely that the small asymmetry in the body profile is sufficient to counteract buoyancy dynamically.

Contrary to fish and dolphins, the penguin's trunk does not participate in thrust production itself. Trunk oscillations during a wing-beat cycle are moderate. Thus the body can be ideally streamlined in shape to be moved through the water with minimum fluid resistance. Recent visualization experiments with life-sized plastic models (Bannasch & Fiebig, 1992) in a smoke-wind tunnel showed a smooth flow around the body, separation occurring, if at all, only in the tail region at lower Reynolds numbers (Oehme & Bannasch, 1989). The fluid resistance of the trunk models representing examples of the three species of the genus *Pygoscelis* was measured in a circulating water tank ($10^5 < Re < 6.3 \times 10^6$). Very low drag coefficients in reference to the frontal area ($0.03 < C_D < 0.06$) were found (Bannasch, in press). Coincidence with the data of Nachtigall & Bilo (1980) was observed at lower Reynolds numbers. For the frontal drag coefficient of the trunk itself these authors presumed a value of ~0.05. Considering that in contrast to my models the body of a live penguin is adaptable to the respective flow conditions, and that the compliant and fine-structured surface of the plumage may be able to produce damping effects, I assumed a realistic C_D-value for the trunk at cruising speed (2-2.5 m s^{-1}) to be $0.03 < C_D < 0.04$ (Bannasch, in press). In a preliminary calculation from energetic studies on living adelie penguins in a swim canal, we derived a drag coefficient of 0.0368 at a swimming speed of 2.2 m s^{-1} (Culik *et al.*, in press). This value is better than the drag coefficient of an ideal spindle ($C_D = 0.04$, Nachtigall & Bilo, 1980) and may explain the exceptional economical use of energy in subaquatic flight of penguins found by Culik *et al.* (in press).

However, one should be careful in extrapolating from the present data to penguin species of other sizes. Depending on their size, as well as on their preferred cruising

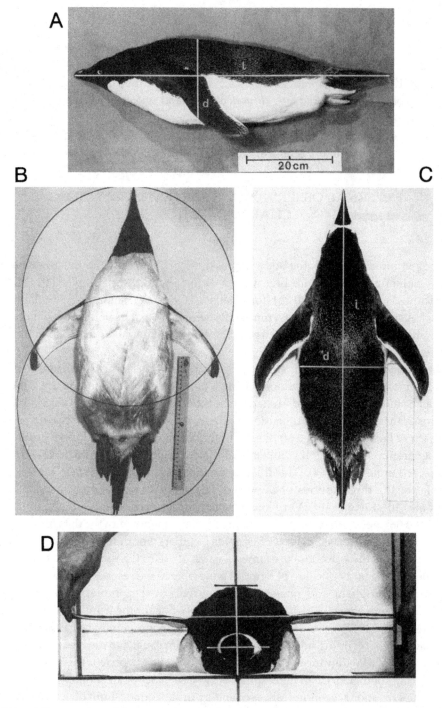

Figure 1. Posture adopted during swimming. (A) *Pygoscelis adeliae*, lateral view. (B) *Pygoscelis papua*, ventral view. (C) *Pygoscelis papua*, dorsal view. (D) *Pygoscelis papua*, frontal view. For description see text.

speed, they all swim at different Reynolds numbers, and therefore the geometry of the flow around the body, and consequently the optimum body shape, cannot be exactly geometrically similar. Some details in the proportions of head, body, tail and wings are quite different in the various species (cf. Livezey, 1989), and should be checked also in the light of hydrodynamic scale effects.

Flight apparatus

Figure 1B shows the ventral view of a gentoo (*Pygoscelis papua*). The body may be divided into three parts of about the same length, the length of the flight apparatus approximately lies within the upper and lower limits of the central section.

It is interesting that in the horizontal plane the caudal position of the wings (gliding posture) is anatomically fixed. In this position, their rounded cranial margins form an arc of a circle. To be fully pressed to the body, the wings must be pronated first. In this posture their tips reach exactly the end of the *sternum*. In the most anterior position of the wings, their long axis is perpendicular to the long axis of the body (Figure 1C).

In the frontal view, bearing in mind the hypothesis of a symmetry of up- and down-strokes, one would expect a change from a 'shoulder-wing plane' favourable in aerial flight, to a construction similar to a 'mid-wing monoplane'. The position of the shoulder joint, however, presents a compromise between these two extreme technical analogies (Figure 1D). Since the wings are never flexed during all phases of the wing-beat cycle, all functions managing the wing movement are centralized in the shoulder joint.

The rotating mechanism is of particular interest because of the high degree of mobility, particularly with respect to supinatory wing rotation. Compared to flying birds, the up-stroke and down-stroke require a high and adjustable angle of attack. The examination of the ability of an abducted wing to be rotated (Figure 2) revealed, however, that pronation is limited at an angle of about -53° (referring to the horizontal plane), whereas the supination was restricted to +39° (mean values of 287 measurements in 19 penguins of 5 species). Apart from an asymmetry, the limited angles for rotation showed that the wings cannot be used for paddling, which would require the wing plane to be orientated perpendicularly to the longitudinal axis of the body.

In the transverse plane, the supinated wing (up-stroke position) could be elevated to an angle of ~+45°. If further raised, it automatically turned around an axis close to its leading margin into the most pronated position (Figure 3). Obviously there is a mechanism in the shoulder joint switching the wing automatically from the up-stroke to the down-stroke position at this point. It was surprising to learn that the axis of wing rotation did not pass through the shoulder joint, but had a more cranial position. No restrictions were found for wing depression.

Wings

The penguin's flipper seems so unlike the wing of a flying bird that at first sight it is difficult to believe that penguins have flying ancestors. The broad discussion in the literature on this topic was reviewed by Bannasch (1986a,b, 1987). Wiman (1905),

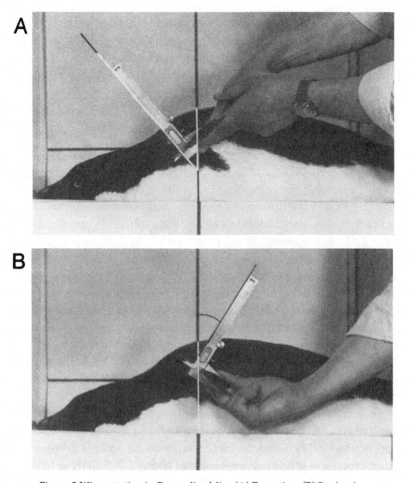

Figure 2.Wing rotation in *Pygoscelis adeliae*. (A) Pronation. (B) Supination.

Simpson (1946) and Storer (1960) have assumed a series of adaptive stages through which the penguin wing may have passed, and the functional-anatomical analysis of the joints by Stegman (1937, 1970) offered proof that the contested wing structure gives clear evidence for airborne ancestry. This general conclusion was further supported by canonical analysis of skin and skeletal measurements (Livezey, 1988, 1989).

Contrary to flying birds, the wings of penguins are short and stiff, the muscles inside are reduced as well as the articulations, the elbow joint is additionally locked by sesamoid bones in the tendon of the triceps, all bones are compact and flattened (cf. Livezey, 1989), the skeleton is covered by firm connective tissue and the numerous feathers are short and stiff. Overall flexibility of the wing is still preserved.

The contours of the wings were documented by self prints on photographic paper. For comparison, they were drawn to the same scale by using the anatomical wing length as a standard reference.

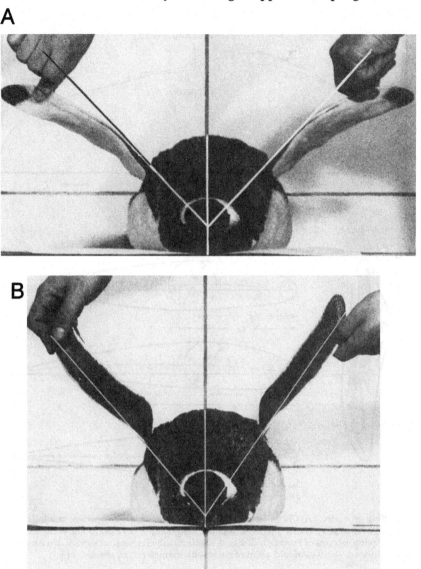

Figure 3. Wing elevation in the transversal plane. (A) In supinated position. (B) In pronated position.

The chord length distribution (Figure 4A) shows the wings of all three pygoscelid species to be relatively slender, with the tip rounded. The gentoo has the most slender wing, the adelie the broadest one with only small differences compared to the chinstrap penguin. The aspect ratio Λ, given by $\Lambda = 2 s^2 / A^*$, where s denotes the free length of the wing (m), and A^* the area of one wing (m²), was 9·07, 8·42, and 8·57 respectively (for details see Bannasch, in press).

Considering the obviously much lower aspect ratio in the wings of little penguins and the much higher one in emperors (an opposite tendency reported by Livezey, 1989,

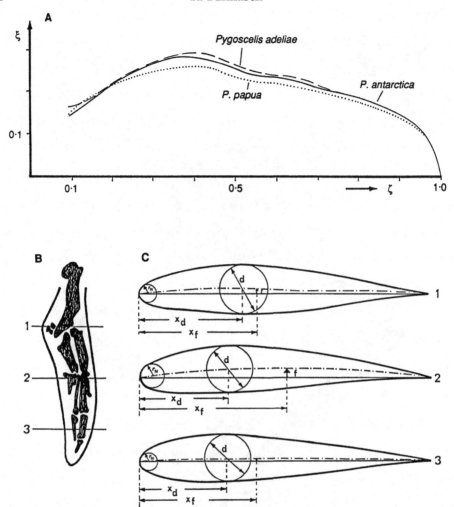

Figure 4. Wing geometry. (A) Chord length distribution (means for each species) in wings drawn to the same scale. x, relative chord length; z, relative wing length. (B) Position of wing sections. (C) Geometry of wing sections of *Pygoscelis adeliae*. d, maximum thickness; f, maximum of mean chamber; rN, nose radius; xd, abscissa of mid point of circle with diameter d; xf, abscissa of f.

was wrong), I assume that there is also an interesting scale effect in wing geometry which needs to be investigated (also in view of probable correlation with the proportions of the trunk). The increase of wing span and aspect ratio clearly points to a reduction of the induced drag with size progression.

The outlines of the wing profiles were measured at three points (Figure 4B) by using the 'Profilkammtechnik' (Nachtigall & Wieser, 1966; Oehme, 1970; Nachtigall & Klimbingat, 1985), and were then enlarged to the same size taking the respective chord length as a standard reference (Figure 4C).

In general the profiles of the three pygoscelid species at the same measuring point may be assumed identical. They are thick throughout and low chambered with rela-

Figure 5. *Pygoscelis papua* after skin removal.

tively rounded noses. The maximum thickness position increases in the distal parts of the wing whereas changes in the thickness ratio over the length of the wing are moderate. Such nearly symmetrical profiles are shown (Prandtl, 1935; Prandtl & Betz, 1935) to yield, in case of negative angles of attack, a relatively high negative lift along with low drag profile. Best performance should be expected with small positive angles of attack (down-stroke).

DISSECTION

After removal of the skin, the dominant large pectoral part becomes visible. The *musculi pectorales* cover nearly four-fifths of the trunk. Two flat tendons at the surface seem to divide the muscle into three parts. The *sternum* ends with an elastic cartilaginous process which is not covered by muscle. Both pectoral muscles are fused together, covering the *carina sterni* completely. From Figure 5 it becomes clear that the mean pulling force of the pectoral muscles is directed more distally than in flying birds (because no weight support is necessary).

The pectoral girdle

The *sternum, coracoidae, furcula* and *scapulae* form the stiff body of the flight apparatus, providing the articulation surface for the *humerus* and the base for origin of the flight muscles (Figure 6). Their function should be considered together. First reports on penguin skeletons (Cuvier, 1800; Meckel, 1828; Reid, 1835; Owen, 1836, 1866; Brandt,

1840; Coues, 1872; Gervais & Alix, 1877) were general and in many cases fragmentary. More complete descriptions were published by Filhol (1882 a,b, 1884, 1885), Watson (1883), Menzbier (1887) and Fürbringer (1888). Later dictionaries considering penguin anatomy (e.g. Gadow, 1893) were mainly based on the description given by Watson. However, the discussion of several apomorphies in the skeletons of penguins was continued (e.g. Pycraft, 1898, 1907; Shufeldt, 1907), and frequently stimulated by new discoveries of fossils (e.g. Wiman, 1905; Lowe, 1933, 1939; Simpson, 1946). In this context, a number of functional aspects has been considered (e.g. Stegman, 1970) but a comprehensive functional-anatomical study of the 'flight apparatus' in penguins is still lacking. Also the comparative osteology of penguins published by Stephan (1979) and the morphometric patterns analysed by Livezey (1989) were focused mainly on zoometric differentiation within the taxon, with only a few measurements and proportions really related to the biomechanical properties (statics and dynamics) of the structures. Special functional studies are required.

Sternum

The *sternum* is long and slender, as in most diving birds. The statement of Kooyman (1975) about a reduced breast bone in penguins was wrong. According to Stephan (1979) its length is 57·7-69·2% of the trunk length. Fürbringer (1888) distinguished between the part connected with the ribs (*costosternum*) and the caudal part (*xiphosternum*). The latter consists of the *trabecula mediana* and the *trabeculae laterales* connected by *membranae incisurarum sterni*. The cartilaginous processes arising from the caudal ends of the *trabeculae laterales* overlap each other, and are fused together forming a plate which, together with the *membranae incisurarum*, support the *trabeculae* to withstand the lateral forces produced by the strong *musculi pectorales*. The springy construction and the elastic cartilaginous plate may also be useful to prevent damage to the *sternum* when the penguin passes through pack ice or makes a crash-landing after jumping out of the water.

Two *processi craniolaterales* arise from the cranial edge of the *costosternum*. These were largest in *Aptenodytes forsteri*. At the *facies musculus sterni* (*facies ventralis s. externa*), a *carina* is well developed. Its pointed *apex cranial* protrudes far beyond the *corpus sterni*. Caudally it nearly reaches the end of the *trabecula mediana*, thus no *planum postcarinale* may be observed. In fledglings I found only the costal part of the *sternum* to be ossified.

Coracoideum

The *coracoidae* are the most massive bones of the pectoral girdle. They are relatively long, and their *facies dorsales* lie in one plane with the lateral edges of the *sternum*. In the central part of the *coracoideum* the *facies dorsalis* is flat, in the caudal part it becomes convex (*impressio musculi sternocoracoidei*). The *facies ventralis* is convex. In penguins the *processus acrocoracoideus* is long. Its cranial part is bent medioventrally and forms the *facies articularis clavicularis*. The *impressio ligamentum acrocoracohumeralis* exists but is not pronounced. The *sulcus musculi supracoracoidei* is large, taking account of the strong tendon of the up-stroke muscle. In penguins, the *processus procoracoideus* is particularly

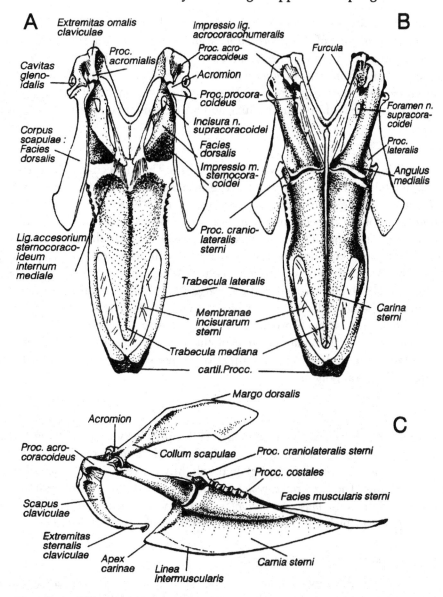

Figure 6. The pectoral girdle. (A) Dorsal view. (B) Ventral view. (C) Lateral view.

strongly developed. Its mediocranial part provides the second surface for the syndesmotic connection with the *furcula* and with the *acromion* of the *scapula*.

The sternal part of the *coracoideum* is broad with a well-developed *processus lateralis*. The *facies articularis sternalis* approximately describes a quarter of a circle. A remarkable variability in the ossification of the coracoidal edges of the *membrana sternocoracoclavicularis* was found. A strengthened part of this membrane (*ligamentum accessorium sternocoracoideum internum mediale*) connects the *procoracoid* with the *angulus medialis sterni*. Partial ossifications mainly arise from the procoracoid and bend to the *corpus coracoidei anterior* of the *impressio musculus sternocoracoidei* and/or propagate along the ligament in its mediocaudal direction. Ossification may also arise from the *margo medialis* and from the *angulus medialis*. In many cases, the *incisura nervorum supracoracoidei* was closed to a foramen. These ossifications are highly irregular, show great variations even within an individual, and are dependent on the age of the bird.

Scapula

The *scapula* lies on the coracoid over its entire width. The connection between the two bones is relatively stiff but not synostotic. The *acromion* is medially connected with the *clavicula*. Lateral to the *scapula*, the *facies articularis humeri* is connected with the corresponding facies of the coracoid by the *ligamentum coracoscapulare interosseum* forming the *cavitas glenoidalis*. The shaft of the *scapula* is relatively slender, the larger caudal part is broad and shovel-like. From the dorsal view the shoulder blades first lie parallel to the vertebral column, then they bend to the sides giving space for the beginning of the *pelvis*. Stephan (1979) emphasized that the *margo dorsalis* has the same position as the dorsal edge of the shoulder blades in recent flying birds. The difference is that in penguins the *scapula* is much broader, overlapping also the *processi uncinati* of the middle thoracic ribs. The enlarged area of the shoulder blades and their position correspond to the strengthening of the up-stroke muscles and the need for a stronger backward-pulling component.

Furcula

The quite massive *claviculae* are fused together, forming the v-shaped springy *furcula*. Their upper parts are laterally flattened, and are broadest at the place of articulation with the acrocoracoid (*scapus claviculae*). In the slightly convex *facies cervicalis*, differentiation is moderate. The *facies coracopectoralis* shows a broad shallow impression deriving from the origin of the *musculus pectoralis*. Both branches bend caudally, but do not reach the *carina*, to which they are connected by the *membrana cristoclavicularis*.

Functional peculiarities of the pectoral girdle

As in other birds, the only real joint exists between the *sternum* and *coracoideum*. The peculiarity in penguins is that it is curved (Figure 7). Marples (1952) has also found this curved shape in fossil penguins, and Simpson (1975) pointed to the need for a functional interpretation of this structure.

The first explanation is the stability to withstand the strong forces of the *musculus pectoralis* as well as that of the enlarged *musculus supracoracoideus*. In relation to the *plana horizontalia*, the coracoid (dorsal surface) and lateral margins of the *sternum* plate are arranged at an angle of ~15° (Figure 6C) which corresponds to the more caudoventral total force of the *musculi pectorales*. In comparison to flying birds the penguin pectoralis is much longer and broader. Therefore its ventral and caudal force components are more pronounced, causing stronger moments at the articulation point. A curved articulation surface may better meet the various configurations and directions of forces (including also action of the extremely strong *musculus supracoracoideus*). The *musculi sternocoracoidei* are comparatively strong. Additionally, in conjunction with the *ligamentum accesorium*, a triangular construction is achieved, preventing (together with *furcula* and *membrana sternocoracoclavicularis*) simultaneously a lateral dislocation of the coracoids.

The second consideration concerns breathing. If the *thorax* is encased by three large bone blades (*sternum* and shoulder blades), how can the bird breathe? The shoulder blades cannot be moved relative to the coracoids because this would change the configuration of the *cavitas glenoidalis*. Elevation and depression of the coracoids relative to the *sternum* are prevented. But the articulations allow some gliding of the coracoids in medioventral and dorsolateral directions which, because of their curved articulation surface, result in rotation of the coracoids around their longitudinal axes. The flexible connection with the *furcula* allows this motion. The rotation of coracoids causes gliding of the shoulder blades along the circumference of the thoracic cage. All muscles originating from the *scapulae* and attached to the *humerus* act to enlarge the distance between the shoulder blades. Consequently, in the *articulatio sternocoracoidalis* supination must be prevented - this is due to the lateral position of the *cavitas glenoidalis*, the cranial and central parts of the *musculus pectoralis* also produce pronating forces. This explains the main mediocranial direction of the muscle and tendon fibres of the *musculi sternocoracoidei* originating from the cranial margin of *processi laterales*. Apart from the forces elevating the coracoids these muscles also pronate them. Consequently they move the *scapulae* towards the vertebral column. Thus they may be considered as synergists of the *musculi rhomboidei* and, in view of the rotating forces at the coracoid, as antagonists of the *musculi pectorales*. Supination of the coracoid is further restricted by the *capsula articularis* and the *ligamentum accessorium sterno-coracoideum internum mediale*,

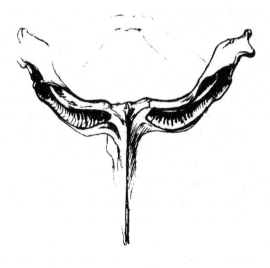

Figure 7. *Sternum* viewed frontally.

and pronation is strongly limited by the *processus craniolateralis sterni*. In general, rotation of the coracoid is considerably small, and does not effect the function of the shoulder joint.

Articulatio humeralis

Morphology

Fürbringer (1888) has described the shoulder joint of birds as a highly developed saddle joint. According to Sy (1936), it provides in principle the same functions as a ball joint. However, apart from gliding, it also involves a certain rolling component, and rotation is limited not by ligaments but primarily by shape configuration.

The shoulder joint consists of three essential parts: *caput humeri, cavitas glenoidalis* and the tendon system (*ligamentum acrocoracohumerale, capsula synovialis* with several structural specializations).

Figure 8 shows that the *caput humeri* is much larger than the shallow *cavitas glenoidalis*. With respect to its oblong semi-oval convex shape, the possibility of a real rotation (as, for example, described for the human shoulder joint) is doubtful. In each moment only a part of the *facies articularis humeri* can articulate with its support. However, a detailed description of the contour of *caput humeri* does not allow a reconstruction of changes of its position during wing rotation. While the caudal outline suggests one type of movement (mainly rotation), the cranial outline suggests a gliding movement. However, only the more central part of the convexity is enclosed by the margins of the *cavitas glenoidalis*.

The difficulties in explaining the joint mechanics on the basis of morphological descriptions of its parts are increased by the complicated shape of the *cavitas glenoidalis*. The nearly flat osseous articulation facets of the *scapula* and the coracoid are inserted with the thick fibrocartilaginous *ligamentum coracoscapulare interosseum*. The elevated margins of this ligament produce the shallow glenoid cavity. The proximate angles between the labiums and the shaft of the coracoid are shown in Figure 8B.

Contrary to the opinion of Sy (1936), emphasizing the flexibility of the floor of the cavity and the change of its shape in response to contact with different aspects of *caput humeri* with which it articulates, I found the *cavitas glenoidalis* of penguins to be relatively invariant. But even if a certain elasticity of the floor of the cavity is granted, it does not play a substantial role in the adjustment of the movement of the *caput humeri*. More important is the fact that this bone is wedged into the lips of the *cavitas glenoidalis*, supporting it at two points. These lips could not be moved apart by the *caput humeri* as Sy (1936) postulated to explain the rotation.

Generally this construction with only two supporting points seems to be too weak to withstand exclusively the forces produced by the strong muscles of the pectoral girdle.

Functional investigations

X-ray examination. X-ray analyses were carried out at the hospital of the Russian Antarctic station Bellingshausen. The aim was to determine the aspects of the *caput humeri* articulated in different postures of the wing.

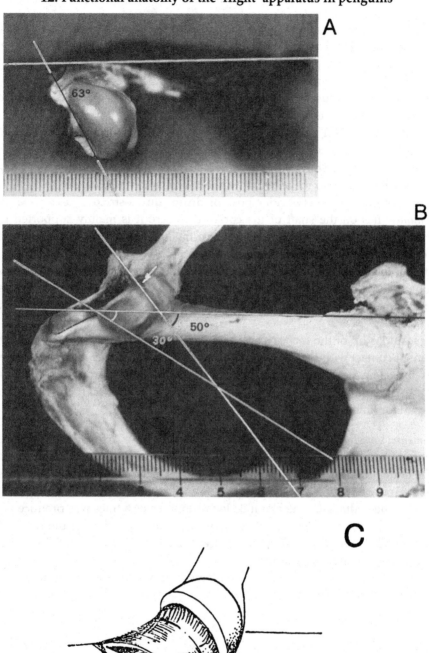

Figure 8. The *articulatio humeralis* in *Pygoscelis papua*. (A) *Caput humeri*, medial view. (B) *Cavitas glenoidalis*, lateral view. (C) Schematic drawing clarifying the shape of the *cavitas glenoidalis*. The various angles are used for the construction of the kinematic model. The contours of the points of attachment of the *ligamentum acrocoracohumerale* are marked.

A gentoo penguin was frozen with one pronated (left side) and one supinated (right side) wing. After taking a picture from the ventral view (Figure 9), the thoracic segment was isolated and a further picture was taken from the frontal view (Figure 10).

Another bird of the same species was dissected so that the shoulder joint of one side of the body (with the muscle, tendon and skeletal system of this side undisturbed) could be examined without interference from the contralateral part of the body. During x-ray examination the lateral stretched wing was turned at first to an extreme pronation (Figure 11A) and then to an extreme supination (Figure 11B).

These x-ray pictures show that during wing rotation the *caput humeri* moves up and down perpendicularly to the *margo lateralis* of the shaft of the coracoid.

In the pronated position (working position during down-stroke), the convexity of the *caput humeri* lies on the shaft of the coracoid, where it is mainly supported by the coracoidal lip of the *cavitas glenoidalis*. Hence the *humerus* can be levered against the shaft of the coracoid by the *musculus pectoralis*.

With supination of the wing, the convex part of the *caput humeri* glides backwards and downwards beneath the shaft of the coracoid. Here it gets support against the elevating force of the *musculus supracoracoideus*, the main wing-raising muscle during the up-stroke.

The scapular lip of the *cavitas glenoidalis* obviously adjusts these movements by its permanent contact with the convex caudal part of the *facies articularis humeri*, preventing at the same time the *caput humeri* from gliding backwards out of the cavity. Therefore the study of the interaction between the scapular lip and the *caput humeri* may be helpful for kinematic considerations.

Experimental investigation of the rotatory movement of the wing. To find out the curvature of the humeral articular facies essential for wing rotation, a fine-tipped pen was inserted into the undisturbed joint through an artificial notch in the scapular lip of the cavity (see arrow in Figure 8B).

Rotating the abducted wing about its longitudinal axis a trace was produced on the *caput humeri*. After isolation the *humerus* was photographed from the medial view (compare with Figure 8A). Using an enlarger the contour of the *caput humeri* and the trace were hand drawn (Figure 12).

The trace could be well-approximated by a sector of a logarithmic spiral line of the common form (in polar co-ordinates):

$$\rho = e^{\left(\frac{\varphi}{\tan\tau} - \frac{c}{2}\right)} \tag{4}$$

where ρ is the length of the radius vector, φ is its angle with the abscissa, τ the (constant) angle between the tangent and the related radius vector, and c is a constant.

The parameters estimated for the curve are $\tau=72\cdot76°$ and $c=3\cdot50$; approximated sector $55° < \varphi < 145°$; P_o marks the centre of the spiral line which coincides with the centre of attachment of the *ligamentum acrocoracohumerale* to the *caput humeri*.

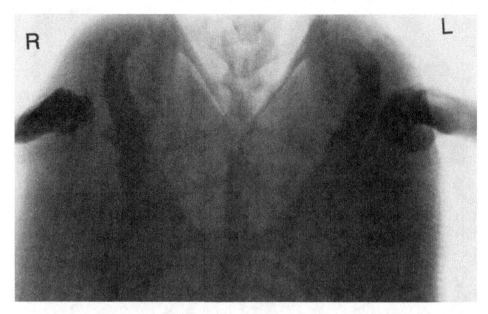

Figure 9. *Pygoscelis papua*, x-ray pictures, ventral view, clarifying the movement of the *caput humeri* in the shoulder joint. L, pronation; R, supination.

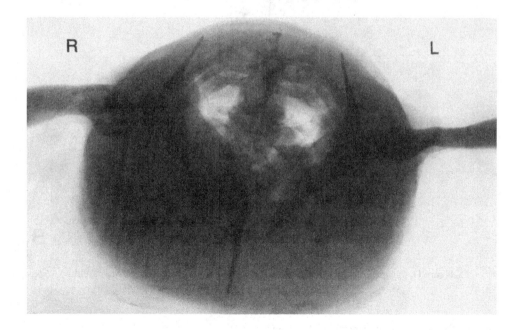

Figure10. *Pygoscelis papua*, x-ray pictures, frontal view, clarifying the movement of the *caput humeri* in the shoulder joint. L, pronation; R, supination.

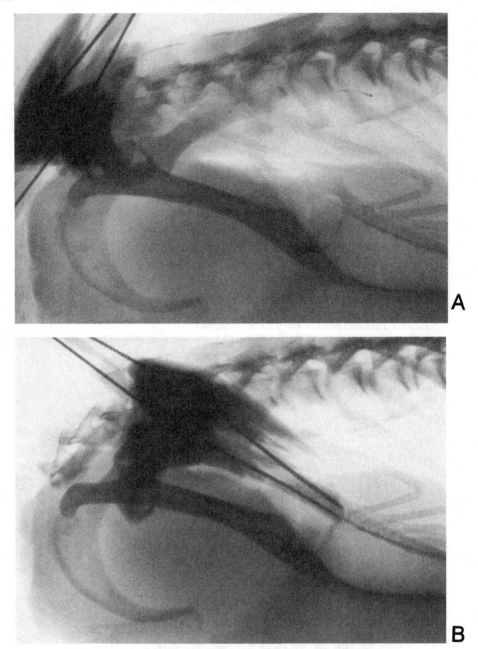

Figure 11. *Pygoscelis papua*, x-ray pictures, lateral view, clarifying the movement of the
caput humeri in the shoulder joint. L, pronation; R, supination.

The model of humerus movement

The wing rotation may be understood only by considering the tendon system of the
joint. Attention must be drawn to the *ligamentum acrocoracohumerale*, the principal

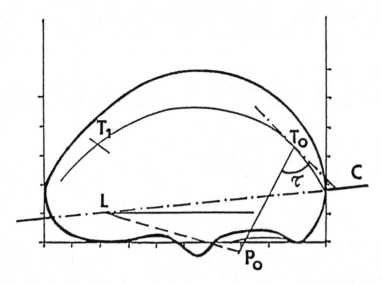

Figure 12. Trace of the movement of the *caput humeri* and its approximation to a section of a logarithmic spiral line. T0 and T1 limits of the section are used for construction of the kinematic model. L, point of attachment of the *ligamentum acrocoracohumerale* to the acrocoracoid; Po, point of its attachment to the *caput humeri*; t, tangential angle of the logarithmic spiral (the tangent coincides with the axis of the scapular lip of the cavity); C, axis of the coracoid.

collateral ligament of the shoulder joint of birds, which is usually independent of the *articulare capsule*. Earlier authors discussed the function of this ligament as limiting the pronation.

In fact this ligament co-ordinates the complex movement of the *caput humeri* in the shoulder joint, permitting at the same time the shifting movement of the bone described above, and, in connection with the scapular lip, prevents it from gliding backwards out of the joint.

The *ligamentum acrocoracohumerale* is stretched in all working positions of the wing, because the thrust forces produced during both the down- and up-strokes have to be transmitted to the body of the bird by backward-pulling muscle components attached to the *humerus*. So the ligament operates like the rigid arm of a lever, and along with the guidance by the lips of the *cavitas glenoidalis*, supports the motion of the *caput humeri* at a further two points.

In this connection the following two peculiarities of the ligament are significant: (1) The point of attachment of the ligament to the *processus acrocoracoideus* of the coracoid allows rotation of the ligament and the related *caput humeri* about an axis through this point which is fixed in relation to the body. (2) The point of attachment of the ligament to the ventral surface of the *caput humeri* allows a further rotation of the *caput humeri* about a second eccentric axis through this point, itself circulating about the first one.

Now the kinematic model illustrating this two-axis eccentric movement of the *caput humeri* can be established (Figure 13). This model demonstrates why the trace (Figure 12) may be closely approximated by a logarithmic spiral line. This function is characterized by a constant angle between its radius vectors and the related tangents. If the tangent is considered as fixed (scapular lip of the *cavitas glenoidalis*), movements of the spiral cause rectilinear movements of its centre (point of attachment of the *ligamentum acrocoracohumerale* to the *caput humeri*).

The deviation of the movement of the spiral centre from a theoretical stated circular movement is negligible if one takes into account the specific features of organic joint construction (e.g. flexibility), the high degree of model abstraction and the relatively small sector of the spiral line covered during wing rotation.

The horizontal line in the first picture approximately marks the level at which the *caput humeri* remains in contact with the convex dorsal surface of the coracoidal lip of the cavity (tangent) during the motion. That it exactly follows the contour shown in the scheme of Figure 8C may be considered to be proof of the kinematic model. The model also coincides with the experimental finding on p. 169 that the main axis for wing rotation does not pass directly through the shoulder joint, but is positioned somewhat cranially. Now it becomes clear that the fixed centre of wing rotation (with respect to the body) is the point of attachment of the *ligamentum acrocoracohumerale* to the *processus acrocoracoideus*.

Functional aspects of the capsula articularis:

The *capsula articularis* enclosing the joint is not as highly differentiated as described in flying birds, e.g. by Sy (1936). Nevertheless, during wing motion several zones of tension may be distinguished, which I named in accordance with the respective intracapsular configurations described by Fürbringer (1888).

Figure 14 clarifies how these substructures of the *capsula articularis* limit the maximum angles of rotation. Pronation and forward gliding of the *caput humeri* in the *cavitas glenoidalis* are limited by the *ligamentum scapulohumerale dorsale*. In the pronated position the *caput humeri* gets additional support at its ventral side by the *ligamentum acrocoracohumerale*. Supination is limited by the *ligamentum coracohumerale dorsale* and the *ligamentum scapulohumerale posticum*. The latter prevents the *caput humeri* from gliding downwards out of the joint.

With elevation of the abducted supinated wing, the caudo-ventral part of the *caput humeri* moves in a more lateral position relative to the articulation point, thereby straining both the *ligamentum acrocoracohumerale* and the caudal part of the *capsula articularis*. Because their length is limited and their projection in the medial plane becomes relatively shorter, the head of the *humerus* must move forwards and upwards in the *cavitas glenoidalis* if wing elevation is continued. Consequently the wing becomes pronated. This seems to explain the mechanism switching the wing automatically from the up-stroke to a down-stroke position at a certain angle of elevation as observed on p. 169. Further evidence for this mechanism can be found in the muscle section.

Conclusions

The clarification of the structures of the shoulder joint in penguins, and the possibilities of performing complicated and effective wing movements, should be borne in mind

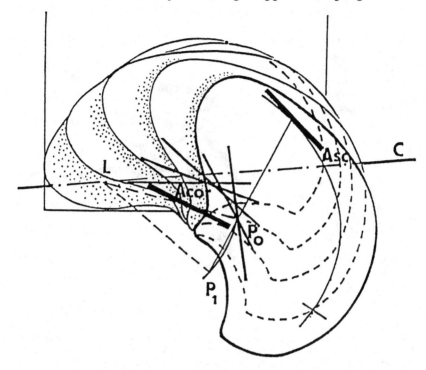

Figure 13. Kinematic model of the two-axis eccentric movement of the *caput humeri* in the shoulder joint. Starting position as for Figure 12. P1, final position of the point of attachment of the *ligamentum acrocoracohumerale* to the *caput humeri*; Asc, axis of the scapular lip of the cavity; Aco, axis of the coracoidal lip (see text).

if the mechanics of wing motion of all birds is discussed. I suppose that in principle the kinematic model of the penguin's shoulder joint may be applied with variations to the shoulder joint of all birds; nevertheless the up-stroke of flying birds is more complicated than that of penguins.

Furthermore, it may be supposed that comparative anatomical studies of recent and fossil penguins, on the basis of the presented kinematic idea, may give the key to an interpretation of several variations in the shape of the humeral bones (in particular of fossil specimens) as functional adaptive stages in the evolution of the locomotor system.

The muscle system

Apart from an increased myoglobin concentration, adaptation of the highly evolved flight apparatus of birds to subaquatic locomotion should be expected to involve mainly a rearrangement of the macrostructures and changes in their respective proportions. The flight muscles of birds utilize fat aerobically as the main energy source, but in the pectoral muscles of many birds there are some fibres which have the ability to use glycogen, with the advantage of a quick mobilization of energy in special situations (start, landing, escape etc.). These fibres also have the capability for brief anaerobic action. One could imagine that an increase of the number of glycogen fibres would

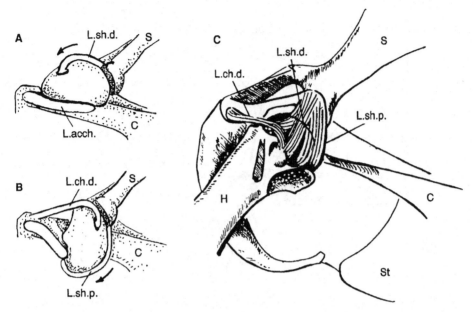

Figure 14. Function of several substructures of the *capsula articularis* in order to limit the maximum angle of (A) pronation and (B) supination. L.sh.d., *ligamentum scapulohumerale dorsale*; L.sh.p., *ligamentum scapulohumerale posticum*; L.ch.d., *ligamentum coracohumerale dorsale*; L.acch., *ligamentum acrocoracohumerale*; C, coracoid; S, *scapula*; H, *humerus*; F, *furcula*; St, *sternum*.

favour maximum diving capacity in penguins. However, single deep anaerobic dives would require lengthy periods of recouperation, prohibiting repeated foraging dives which are, obviously, more efficient and allow a greater percentage of time to be spent feeding under water (Baldwin, 1988). In my histological studies, analysing fibre morphology (Bannasch, 1986b), I could not find evidence for an increased occurrence of glycogen fibres. Also, the study of the ultrastructure and biochemistry carried out by Mill & Baldwin (1983) led to the conclusion that the basic metabolism of the flight muscles in little penguins works aerobically, and that their capacity for anaereobic glycogenolysis is strongly limited. But a later comparative examination of the muscle biochemistry in seven species of penguins revealed an increasing anaerobic capacity of the muscles used to power swimming in the following arrangement: little<royal and rockhopper<gentoo<adelie, emperor and king (Baldwin, 1988). However, our field observations (Culik *et al.*, in press) confirmed (as agreed also by Baldwin) that the majority of routine natural dives are aerobic, with surfacing before oxygen stores are exhausted.

The anatomy of the muscle system in penguins was well described by earlier authors (Meckel, 1828; Schöpss, 1829; Reid, 1835; Garrod, 1873; Gervais & Alix, 1877; Filhol, 1885; Watson, 1883; Fürbringer, 1888, 1902), and was recently reviewed by Schreiweis (1982) and Bannasch (1986b, 1987). Thus I can concentrate on some functional aspects. Besides the preponderant thrust-producing (flapping) muscles (e.g. *musculus pectoralis* and *musculus supracoracoideus*) which include rotating components that tend to turn the wing in an extreme pronated or supinated position (respectively), another group of muscles can be distinguished. These muscles control the position of the *humerus* in the shoulder joint, adjusting in this way the hydrodynamic angle of attack according to the flow conditions (cf. Bannasch, 1986b, 1987). In this paper I will concentrate only on the main flapping muscles.

Musculus pectoralis

The main muscle of the down-stroke - the *musculus pectoralis* - is by far the largest muscle of the penguin. However, the impression of its enormous mass given by Figure 5 may be partly misleading owing to the enlarged *musculus supracoracoideus* which lies beneath it. In fact, the mass ratio of both *musculi pectorales* and the body was between 1:8·61 and 1:5·04 in *Pygoscelis antarctica* no. 6 and *Aptenodytes forsteri* respectively), and is therefore within the range of variations found in flying birds (Legal & Reichel, 1880; Magnan, 1922; George & Berger, 1966; Oehme *et al.*, 1977; Bannasch, 1986b). However, because of the compact construction of the penguin skeleton (counteracting buoyancy) and the large amount of fat accumulated, the comparison of this mass ratio to that of flying birds is of limited value. It should also be noted that in penguins this muscle need not compensate the body weight, but can be used for thrust generation only.

I found it remarkable that in its origin the powerful *musculus pectoralis* is mainly displaced from the *carina sterni* by the *musculus supracoracoideus*. Only a small area of the *carina* is left to the *pectoralis* (see *linea intermusculus* in Figure 6C). Somewhat similar to bats (Norberg, 1970, 1972) most of the ventral fibres originate from a fascie deriving from the *carina*. The cranial parts originate from the *membrana sternoclavicularis* (including *ligamentum cristoclaviculare*), the *facies coracopectoralis* of the *furcula* and the ventral margins of the *processus acrocoracoideus*. The caudal parts derive from the lateral margin of the *trabecula mediana*, the *membrana incisurarum sterni* and the *facies ventralis* of the *trabecula lateralis*. By the broad parasternal fascie, deriving from the *trabecula lateralis*, a most caudal direction of the lateral part of the pectoral muscle is achieved. The tendon of this muscle is highly differentiated and inserts onto the *crista pectoralis* (*seu lateralis*) and the linear depression proximal at the shaft to the *humerus*. A small *pars propatagialis* (not described by Schreiweis, 1982) is developed in the pygoscelid penguins.

Despite the somewhat diverging direction of the fibres deep in the muscle, three main pulling components may be distinguished in the muscle. The mass-ratio of the corresponding muscle portions may help to estimate the relative strength of the respective forces (Figure 15).

The *pars propatagialis* (1·5-1·8% of total muscle mass) pulls the wing forwards, and may be active in synergetic conjunction with the *musculus propatagialis* during gliding, when the wing produces only profile drag. During the beating cycle at the upper dead centre the wing experiences only drag forces. The whole cranial part of the *musculus pectoralis* (15%) may be involved in preventing its backward motion relative to the body and in simultaneously starting the down-stroke. The main downward pulling force is developed by the central part of the muscle (~50%). But as soon as the wing produces thrust it tends to move forwards relative to the shoulder joint. Therefore the total

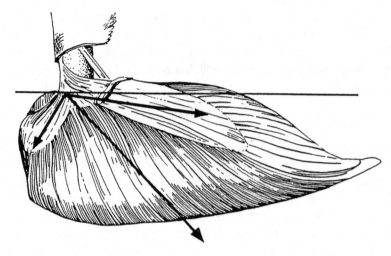

Figure 15. *Musculus pectoralis*. The arrows show the direction of forces produced by different parts of the muscle. The length of the arrows indicates the mass of the respective muscle parts.

muscle force should follow a more caudal orientation. The lateral part (~35%) of the pectoral muscle also becomes involved in transmitting the thrust forces to the body. Like Sy (1936), I presume that during the down-stroke a kind of contraction wave must go through the *musculus pectoralis*. However, the innervation of the muscle with respect to its antagonistic parts needs to be investigated.

As stated above, because of the cranial insertion of its tendon to the *humerus*, the *musculus pectoralis* acts as a wing pronator. However, it is important to mention that in a supinated position the wing's point of attachment lies dorsal to the articulating points of the *caput humeri* with the *cavitas glenoidalis*. In this wing position, the lateral part of the *musculus pectoralis* produces a supinatory component. Its supinating force may be relatively small but is essential if the balance of forces during the up-stroke is considered.

Another interesting functional aspect concerns the participation of the lateral part of the *musculus pectoralis* in the mechanism switching the wing automatically from a supinated to a pronated position at the end of the up-stroke. During wing elevation, its point of attachment at the *humerus* moves in a dorso-medial direction, whereas the ventral surface of the *caput humeri* and the *intumescentia* encapsulating the fossa tricipitalis move outwards. Finally, at an elevation angle of ~+45°, the flat tendon of the lateral part of the *musculus pectoralis* has to pass around the *intumescentia*. If stretched now, it prevents further distal movement of the ventral part of the *caput humeri* and, apart from a pronating force, it also produces a pressure which pushes the *caput humeri* upwards with respect to the cavity (down-stroke position).

Up-stroke muscles

The most powerful muscle of the up-stroke system is the *musculus supacoracoideus*. In penguins the *musculus deltoideus*, the other muscle involved in wing elevation during aerial flight of other birds, is extremely reduced.

As in flying birds, the main up-stroke muscle, the *musculus supracoracoideus* lies on the ventral side of the *thorax*, originating from *facies musculus sterni* (particularly from the *trabecula mediana*, the *membrana incisurarum sterni*, the cranial part of the *trabecula lateralis*, the *corpus sterni* and the *carina*), the basal edge of the coracoid and the *membrana sternocoracoclavicularis* (Figure 16). Its strong, broad tendon moves through the *canalis triosseum* and inserts finally onto the dorsal surface of the *humerus*. This crane-like construction for wing elevation is common to all birds. The stretching of the tendon of the *musculus supracoracoideus* first causes a downward movement of the *caput humeri* with respect to the *cavitas glenoidalis* (wing supination). Thereafter the muscle can develop its wing elevating force by levering the *humerus* against the coracoid. The ventral position of the main up-stroke muscle is the reason for the compromise between 'shoulder-wing plane' and 'mid-wing monoplane'.

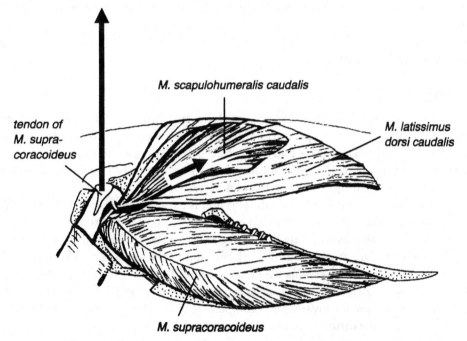

tendon of
M. supra-
coracoideus

M. scapulohumeralis caudalis

M. latissimus
dorsi caudalis

M. supracoracoideus

Figure 16. The main muscles used in the up-stroke. Arrows as in Figure 15.

The peculiarity of penguins is the enormous size of this muscle. In comparison with the *musculus pectoralis*, I found a mean mass-ratio of 1:2·02 in the species I studied (ranging from 1:1·98 in *P. antarctica* to 1:2·27 in *A. forsteri*). In flying birds, similar mass-ratios are found only in hummingbirds. For the latter, a respective mass-ratio of 1:1·7 was reported by Stolpe & Zimmer (1939), whereas Hartman (1961, in King & King, 1979) has found a mass-ratio of 1:2 in this taxon. Both groups of birds produce aerodynamic force also during the up-stroke. It is interesting that Mill & Baldwin (1983) also found some similarities between penguins and hummingbirds in the ultrastructure and biochemistry of their muscles.

Considering the different structures of the muscles, and in particular the pennate architecture of the *musculus supracoracoideus*, the mass ratio *sensu stricto* does not allow conclusions about the respective forces developed by these muscles. A more appropriate method, based on a comparison of cross-sections perpendicular to the muscle fibres, is proposed by Alexander (1968). The application of this method to penguin muscles may be a topic for future research. At the moment a mass-ratio consideration may be useful as a preliminary observation.

Functionally it is important that the *musculus supracoracoideus* can only elevate the wing. The thrust forces thereby produced must be transmitted to the body by other muscles.

The *musculus scapulohumeralis caudalis* and the *musculus latissimus dorsi* (Figure 16) should manage this function. Apart from an elevating component they also pull the wing backwards. The lateral broadening of the *scapula* results in a more caudal orientation of the *musculus scapulohumeralis* (~40° with respect to the long axis of the body). However, the mass of both muscles is too small, only one-fifth of the lateral portion of the *musculus pectoralis*, for example, and therefore exclusively produces backward forces during the down-stroke.

Balance of forces

Based on muscle mass-ratios, in Figure 17 I illustrate and compare the strength and directions of the main forces produced during both stroke phases. The hydrodynamic lift force (L) produced by the flapping wing can be divided into thrust (T) and a positive or negative vertical force (V) during the down- and up-strokes, respectively. These forces must have their opposition in the muscle system.

For the down-stroke (Figure 17A) the solution is easy. Strength and direction of the total force of the *musculus pectoralis* (Fp) exactly correspond to the resultant hydrodynamic force generated by the wing. If a similar configuration of hydrodynamic forces is considered for the up-stroke (Figure 17B), the sum of the vertical components of the forces produced by the *musculus supracoracoideus* (Fs), the *musculus scapulohumeralis caudalis* and the *musculus latissimus dorsi* together (Fd) may well balance the negative vertical force. But the backward force component of the latter two muscles is far too small to transmit the thrust component to the body.

There are two different conclusions possible. Either the up-stroke does not produce considerable thrust forces, which would explain some forward gliding of the wing during the up-stroke, but contradicts the kinematic pictures showing some symmetry of both stroke phases, or the other solution is to balance the thrusting force during the up-stroke by involving the lateral part of the *musculus pectoralis*. The consequence of this assumption would be that the lateral part of the pectoral muscle must be active in both phases of the wing-beat cycle. Thus, in penguins, it is not only antagonistic to the cranial part of the same muscle but it acts also as a synergist of the wing elevating system. Its supinatory action in the up-stroke position of the *caput humeri*, as well as its participation in the mechanism of wing rotation at the end of the up-stroke, can be interpreted in favour of this theory which, however, requires electrophysiological experiments for its confirmation.

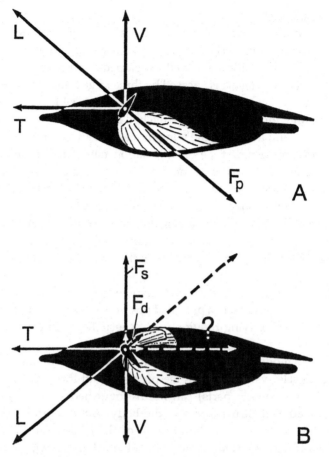

Figure 17. Scheme of balance of forces during (A) the down-stroke and (B) the up-stroke (see text). Assuming symmetry during both stroke phases, the question is, which muscles can transmit the thrust force to the body during the up-stroke?

GENERAL CONCLUSIONS

1. The penguin body seems to be ideally adapted to aquatic life in view of its streamlined shape and specific gravity. In respect to diving capacity, size progression in penguins is not limited by the surface:volume ratio. However, considerable scale effects can be observed in the proportions of the body and the flipper-like wings, as well as in the kinematics of wing motion. As yet, scale effects are ill-studied in penguins.

2. In underwater swimming the body weight is compensated by hydrostatic buoyancy; hence, the strong flight apparatus is used only for thrust production. The wings act as hydrofoils. The mechanism of force generation is, in principle, similar to that of flying birds, but contrary to fast flight in air, mainly thrust is produced, and thrust production occurs also during the up-stroke; consequently, nearly symmetrical wing profiles are developed.

3. In principle, the flight apparatus of penguins consists of the same elements as the original flight apparatus in birds. But the structures have been modified gradually as an adaptation to subaquatic flight. The skeleton of the pectoral girdle became much longer (mainly due to larger coracoids and a well-developed *xiphosternum*), and the shoulder blades were enlarged. The *thorax* is encased by three large bone blades. But a modification of the *articulatio sternocoracoidalis* makes breathing possible.

4. As the penguin wings operate fully stretched in all phases of the beating cycle, all functions managing and adjusting the wing movement are centralized in the shoulder joint, which represents a new type of articulation never previously described. The 'rotation' of the head of the *humerus* in the shoulder joint can be described as a two-axis-eccentric motion, conducted by the *ligamentum acrocoracohumerale* and an articulation surface of the *caput humeri* shaped like a logarithmic spiral.

5. I suppose that, in principle, the kinematic model of the penguin's shoulder joint may be applied, with variations, also to the shoulder joint of all birds, although the up-stroke of flying birds is more complicated than that in penguins. Comparative studies are required.

6. In penguins the joint construction also involves a mechanism switching the wing automatically from a supinated to a pronated position at the end of the up-stroke (angle of wing elevation ~+45°). Wing rotation is not specially automated at the end of the down-stroke.

7. The kinematic model of the shoulder joint leads to a new interpretation of the function of the muscle system. Besides the preponderant thrust-producing (flapping) muscles, muscles (or muscle parts) can be distinguished which are managing the transfer of the produced thrust forces to the body. Another group of muscles (not considered in this paper) control the position of the *humerus* in the shoulder joint, adjusting in this way the hydrodynamic angle of attack of the wing corresponding to the respective flow conditions.

8. Compared to flying birds, the main down-stroke muscle, the *musculus pectoralis*, is enlarged and more differentiated. The caudal and lateral parts are strengthened and rearranged to transmit to the body the thrust forces generated by the flapping wings. Depending on the position of the *humerus*, the lateral part tends to rotate the wing in opposite directions.

9. The *musculus supracoracoideus* is extremely strong in penguins. However, it can only elevate the wing. The backward pulling muscles of the up-stroke system are too weak to balance the thrust forces which may be produced when the wing is elevated by the strong *musculus supracoracoideus*. The lateral part of the *musculus pectoralis* must be involved to manage the transfer of thrust forces to the body.

10. The functional-anatomical consideration supports the impression obtained from kinematic pictures of penguin swimming that the down-stroke has greater reserves for thrust generation than the up-stroke, which may be important for acceleration and fast swimming, and that symmetry of both stroke phases may be restricted to 'economic' swimming with medium power.

Chapter 13

Energy conservation by formation swimming: metabolic evidence from ducklings

FRANK E. FISH

Department of Biology, West Chester University, West Chester, PA 19383, USA

Formation movement, particularly during swimming and flying, has been hypothesized to reduce an individual animal's energy expenditure. Although a number of aero- and hydrodynamic models have been proposed to estimate energy savings of animals travelling in formation, little empirical data are available that test this presumption. An examination of the metabolic energetics of formation swimming was undertaken with mallard ducklings (*Anas platyrhynchos*) trained to swim behind a decoy in a flow tank. Oxygen consumption was measured for ducklings in clutches of one, two or four individuals, with the decoy in the water or suspended above the surface. Metabolic rate per individual duckling decreased with increasing clutch size by 7·8-43·5%. Following in the wake of a decoy significantly decreased metabolic swimming effort by ducklings and this effect was most pronounced at three days of age and in small clutches. These findings suggest that the flow pattern generated by the formation and in the wake of an adult reduces the resistance of the water to locomotion and allows an individual duckling to conserve energy.

INTRODUCTION

Locomotion is an energetically costly activity that may comprise a significant component of an animal's overall energy budget (Weihs & Webb, 1983; Fish, 1992). Therefore it is advantageous for animals to use locomotor strategies which minimize energetic expenditure. Such strategies include gait transition, intermittent locomotion, soaring, tidal stream transport, wave riding, submerged swimming, porpoising and formation movement (Cone, 1962; Lissaman & Schollenberger, 1970; Pennycuick,

Maddock, L., Bone, Q. & Rayner, J.M.V. (ed.). *Mechanics and Physiology of Animal Swimming*.
© 1994. Cambridge University Press.

1972; Weihs, 1973, 1974, 1978; Au & Weihs, 1980; Hoyt & Taylor, 1981; Williams, 1989; Fish *et al.*, 1991; Williams *et al.*, 1992).

Formation movement by animals has been hypothesized to reduce energy expenditure and enhance locomotor performance of individuals during walking (Fancy & White, 1985), swimming (Weihs, 1973; Breder, 1976), and flying (Lissaman & Schollenberger, 1970; Hummel, 1983). Saving energy by moving in formation has been suggested as particularly important for animals that migrate by swimming or flying over considerable distances. Formation swimmers or flyers influence the flow of water or air around adjacent individuals in formations, thereby presumably reducing drag with a concomitant decrease in the overall energy cost of locomotion.

As a mechanism for energetic reduction of locomotor effort, formation movement is accepted generally for automotive and cycling competitions, which use the techniques of 'drafting' or 'slipstreaming'. Trailing cyclists in a pace line experience a 38% reduction in wind resistance and 35% reduction in power output (Kyle, 1979). Reduced metabolic effort of 6·5% has even been measured in the trailing human runners (Pugh, 1971).

Energy savings have been difficult to measure for animal locomotion. The three-dimensional complexity and large size of polarized animal formations, such as fish schools and V formations of geese, have deterred metabolic experimentation to evaluate hypotheses of energy savings by formation movement (Breder, 1976; Shaw, 1978; Hummel, 1983).

Vorticity and relative velocity

The physical basis for energy savings with formation flying and swimming is that as a body moves through a fluid it distorts the velocity field around the body and in its wake. The wake consists of a trail of alternating vortices which are regularly arranged as two staggered rows (Prandtl & Tietjens, 1934; Vogel, 1981; Weihs & Webb, 1983). The vorticity transports momentum within the fluid and affects the velocity profile. Each vortex induces velocities in the surrounding fluid in the same direction as the rotation of the vortex (Weihs & Webb, 1983). The strength of the induced velocity is inversely proportional to the distance from the vortex. Changes in the velocity profile of the fluid influence the relative velocity of trailing bodies in the wake (Breder, 1965, 1976; Belyayev & Zuyev, 1969; Weihs, 1973, 1975). Relative velocity is calculated as the difference between the trailing body's velocity and the mean velocity induced by the vortices shed from the leading body (Weihs, 1973). If a trailing body is moving at a given velocity, and oriented in the same direction as the mean velocity of the vorticity shed into the wake by a leading body, the trailing body will experience a reduction of its relative velocity. Because drag is directly proportional to the velocity squared, a decrease in the relative velocity can decrease drag. A lower drag would require a reduced energy expenditure to generate thrust, which is equal to the drag for steady swimming.

The benefit of travelling in the vortex trail of another body depends on the relative position of the bodies and the mechanism of vortex generation (Figure 1). An oscillating body, such as a swimming fish, generates a thrust-type vortex trail (Figure 1A;

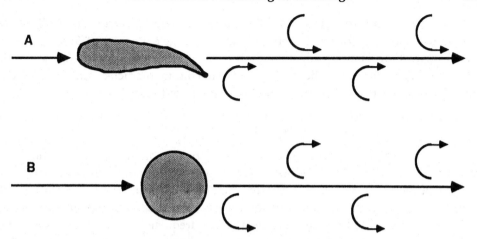

Figure 1. Two-dimensional flow patterns generated in the wake of a body. Arrows indicate direction of flow relative to the body. The thrust-type vortex trail (A) is actively produced by the periodic lateral oscillations of a fish (Weihs, 1972). The drag-type wake of a rigid body (B) has a pattern of vortices that rotate in the direction opposite to the thrust-type vortex trail.

Weihs, 1972). Owing to the rotation of the vortices, a low relative velocity and high energetic advantage are experienced by a fish swimming diagonally behind another fish (Weihs, 1973). Conversely between Reynolds numbers of 4×10^1 and 2×10^5, the rotation of vortices induced by a rigid body is the reverse of vortices generated by lateral oscillations of fish (Figure 1B; Prandtl & Tietjens, 1934; Hoerner, 1965; Weihs, 1975; Vogel, 1981). The flow regime in this drag wake would produce a low relative velocity directly behind a rigid body. In this case animals moving in a single-file formation will experience reduced drag and energy expense. In addition, the vortex pattern behind a rigid body is such that momentum flows towards the body (Rayner, 1985), and a trailing body can be pulled along as momentum is transferred to it from the water. Young animals swimming in the drag wake of a larger adult will benefit particularly from this effect.

Predictive Models

Previous analyses of energy savings by formation movement in animals have focused on aero-and hydrodynamic models (Lissaman & Shollenberger, 1970; Weihs, 1973, 1975; Higdon & Corrsin, 1978; Badgerow & Hainsworth, 1981; Hummel, 1983). These models predict significant energy savings when animals are arranged in appropriate formations. Lissaman & Shollenberger (1970) estimated a 71% saving of induced power for birds flying in V formations.

A three-dimensional, inviscid, hydrodynamic model for fish schools developed by Weihs (1973, 1975) estimated reductions of swimming effort by a factor of five for fish swimming in diamond-shaped formations. Trailing fish in these formations take advantage of vortex trails produced by leading fish. The relative velocity of the trailing fish is estimated at 40-50% of the free stream velocity.

Observations on bird flocks and fish schools indicate, however, that individuals often deviate from the optimal configuration (Partridge & Pitcher, 1979; Badgerow & Hainsworth, 1981; Hainsworth, 1987; Fish et al., 1991). Thus for animals moving in groups, energy savings may be lower than the maximum predicted, and the energetic cost of travel may even be greater than for a solitary individual.

Experimental measurements on formations

Little empirical data are available that test the presumption of energy economy by formation swimming. Fish in schools were reported to swim 2-6 times longer than single fish (Belyayev & Zuyev, 1969). Examinations of swimming kinematics of fish schools indicate energy savings by increasing coast times during burst-and-coast swimming (Fish, et al., 1991) and reduced tail-beat frequency (Fields, 1990). Abrahams & Colgan (1985) measured respiratory rates of schools of three fish swimming in a water current of 0·07 m s⁻¹. Fish were tested as a school and in groups of two separated by a clear partition. A 13% reduction in respiratory rate of the school was found compared to the sum for individuals. However, only schools of large individuals (6 cm length) demonstrated measurable energy savings and the small diameter (5 cm) of the test chamber may have introduced errors due to wall effects.

Queues of spiny lobsters (*Panulirus argus*) in water have been shown to sustain less drag per individual than a single lobster travelling at the same speed (Bill & Herrnkind, 1976). The reduction in energetic cost per individual in a queue was a direct function of queue size. Queues composed of large numbers of lobsters had lower drag per individual than queues of smaller numbers. Drag reduction for individual lobsters was suggested to be important in conserving energy during migration when these animals form queues.

Formation swimming by ducklings

The paucity of empirical data to support or refute the hypothesis of energy conservation through formation movement in animals is a result partly of the cumbersome size of the formation and partly of uncontrolled and inconsistent positioning of individuals in the formation. Energetic measurements have been more forthcoming from studies of line formations in which a rigid body is responsible for the vorticity pattern (Bill & Herrnkind, 1976; Pugh, 1971; Kyle, 1979). Mallard ducklings are well suited for metabolic examination of formation swimming owing to their habit of swimming in single file. These ducklings display a following response due to imprinting on their mother (Hess, 1959; Dyer & Gottlieb, 1990; Bolhuis, 1991) and are capable of being led to water within 12 h of hatching (Bellrose, 1976). A fortuitous consequence of the single-file formation would be a reduction in metabolic effort during swimming, due to a decrease in their relative velocities associated with the wake and pattern of vorticity shed by the leading bodies.

One-day-old ducklings were imprinted on a female mallard duck decoy. The duck-lings were trained daily to swim for 20-30 min at 0·3 m s⁻¹ behind the decoy in the working section (1·2x0·6x0·44 m) of a recirculating water channel (Vogel & LaBarbera,

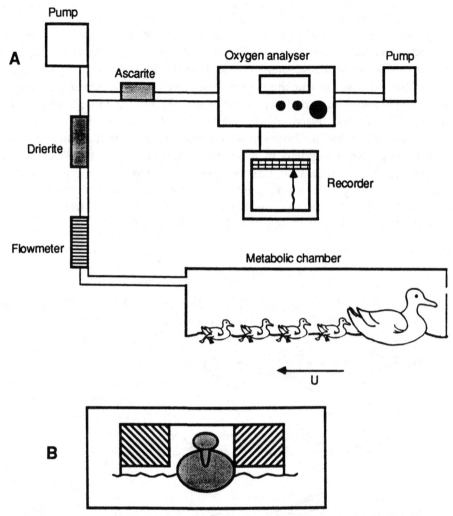

Figure 2. (A) Diagram of oxygen analysis system. (B) Insert shows cross-sectional view of metabolic chamber with decoy. Filled area represent dead-air spaces. Water velocity was controlled by two electric motors (Minn Kota 65MX). Water velocity was checked with either a mechanical propeller flow meter (KAHLSICO Model 005WB138) or electromagnetic flow meter (Marsh-McBirney Model 201). Mean water temperature was 21·6°C ±1·0 (SD).

1978). The upstream end of the working section was bounded by a plastic grid to reduce turbulence in the flow.

A Lucite metabolic chamber was fitted over the working section (Figure 2). The chamber provided a large enough space to allow ducklings to organize into formations and avoid interference with flow and wave patterns generated by the animals. Oxygen-content of dry, carbon-dioxide-free air flowing continuously through the chamber (3-10 l min^{-1}) was monitored with an oxygen analyser (Ametek S-3A/I; Figure 2). Metabolic rate (MR) was measured from oxygen consumption at STP calculated using equations for open circuit respirometry (Hill, 1972) and was expressed in kcal using a caloric conversion factor of 4·8 kcal l^{-1}.

The position of the ducklings in formation and with respect to the decoy was monitored with a video camera. To videotape a dorsal view of the ducklings in the water channel, a mirror was suspended at a 45° angle above the working section.

Twelve groups of seven ducklings were examined. After imprinting, each group was subdivided into three experimental clutches of one, two, and four ducklings. Each experimental clutch was tested at 3, 7, and 14 days of age. The effect of the decoy's wake was determined by ducklings swimming with the decoy in the water or with the decoy raised approximately 0·01 m above the water surface. In the latter position, ducklings could maintain the visual cue of the decoy, but would not experience a wake generated by the decoy. Over a 2-d period, each clutch was tested with the decoy in the up or down position. The order of testing was assigned randomly. Mean masses for 3-, 7-, and 14-d-old ducklings were 0·054 kg ±0·009 (SD), 0·109 ±0·017 kg, and 0·243 ±0·050 kg, respectively.

Metabolic results of ducklings in formation

Ducklings readily followed the decoy when swimming in the water channel (Figure 3). The Reynolds numbers for 3-, 7-, and 14-d-old ducklings, and the decoy were

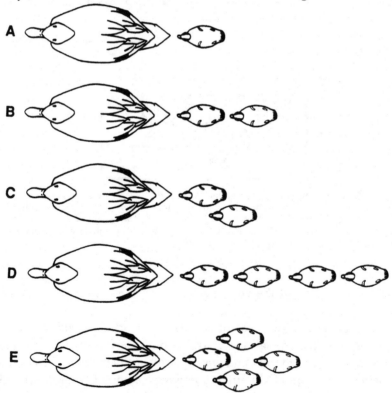

Figure 3. Typical duckling formations for clutches of one, two, and four individuals. Single-file formations are shown in (A), (B) and (D); echelon formation is shown in (C); diamond formation is shown in (E).

$2\cdot15\text{x}10^4$, $2\cdot82\text{x}10^4$, $3\cdot60\text{x}10^4$, and $8\cdot94\text{x}10^4$, respectively, based on waterline length. Solitary or leading ducklings maintained an average position of $0\cdot25$ body-lengths behind the decoy. This distance was maintained regardless of whether the decoy was in or out of the water. Pairs of ducklings were in formations where they swam abreast, one behind the other, or one lateral and slightly behind the other. Clutches of four ducklings typically swam in single-file lines or in diamond-shaped formations. Inter-

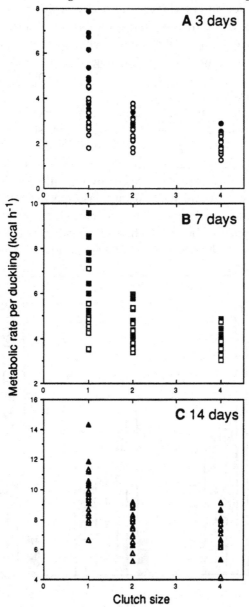

Figure 4. Metabolic rates of ducklings in clutches of one, two, and four individuals at ages of (A) 3 d, (B) 7 d and (C) 14 d, while swimming downstream of a decoy. Closed symbols represent clutches following a decoy suspended above the water surface; open symbols represent clutches following a decoy that is producing a wake in the water.

duckling distances were small and averaged 0·1 body lengths. Ducklings would reposition themselves frequently throughout the testing period.

Metabolic data showed increasing energy economy with increasing clutch size and with influence of the decoy's wake (Figure 4). Significant differences (ANOVA, $P<0.001$) were found with respect to age, decoy position, clutch size, and interaction of decoy position with clutch size. Metabolic rate per duckling decreased by 7·8-43·5% with increasing clutch size for all age classes and decoy positions. The best performance with regard to energy economy was found for 3-d-old ducklings. These ducklings displayed a 62·8% decrease in metabolic effort when swimming in a clutch of four in the decoy's wake compared to a solitary duckling without the decoy's wake. Duckling clutches of one were significantly different (SNK, $P<0.05$) from all other clutches regardless of age or decoy position.

Reduced metabolic effort due to the influence of the decoy's wake was most important in the youngest ducklings. Solitary 3-d-old ducklings showed a 37·7% decrease in MR when swimming in the decoy's wake as opposed to 27·6 and 15·7% decreases for solitary 7- and 14-d-old ducklings, respectively. However as clutch size increased, the influence of the decoy was less pronounced, showing only 7·5-13·1% decrease in MR.

Equivalent results were found when using mass-specific MRs, which were calculated as the clutch MR divided by the total mass of the clutch (Figure 5). The swimming ducklings had mass-specific MRs 2·0-6·6 times the resting MR over the range of ages. Significant differences (ANOVA, $P<0.005$) were found with respect to age, decoy

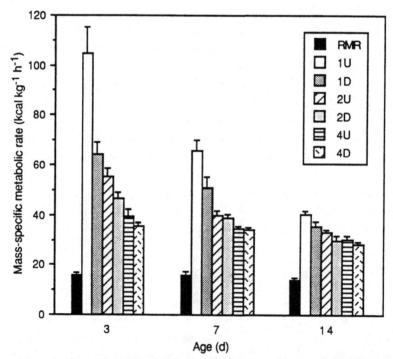

Figure 5. Mass-specific metabolic rates (±SE) of ducklings at ages of 3, 7, and 14 d. Combinations of clutch size (1, 2, and 4 ducklings) and decoy position (U, up above the water; D, down in the water) are shown in the insert. Resting mass-specific metabolic rates are indicated by RMR in the insert.

position, clutch size, and all interactions. Ducklings following in the decoy's wake realized a saving of 20% (SNK, $P<0.05$; all clutch sizes and ages combined) compared to ducklings swimming with the decoy in the raised position. With all ages and decoy positions combined, mass-specific MR decreased by 33% between clutches of one and two ducklings, and by 18% between clutches of two and four ducklings (SNK, $P<0.05$).

Cost of transport

The importance of energy economy by formation swimming for young ducklings is apparent when the cost of transport (CT) is examined. The cost of transport (CT) was calculated from the mass-specific metabolic rate divided by swimming velocity (Schmidt-Nielsen, 1972; Tucker, 1975). The 3-, 7-, and 14-d-old ducklings swimming in four-duckling clutches in the decoy's wake reduced CT by 66·3, 49·6, and 32·1%, respectively (Figure 6). Despite the large reductions in CT, ducklings were found to have the highest CT for any vertebrate swimmer (Schmidt-Nielsen, 1972; Tucker, 1975; Fish, 1992). Solitary 3-d-old ducklings had a CT 88·8 times greater than the minimum CT of a fish of equivalent size, and 17 times greater than an adult mallard duck (Figure 7). These high costs for young ducklings are due probably to scale effects associated with (a) their high maintenance costs of homeothermy in water, (b) their increased drag at the water surface from wave formation, and (c) the inefficiency of their paddling mode (Fish, 1982, 1992; Baudinette & Gill, 1985).

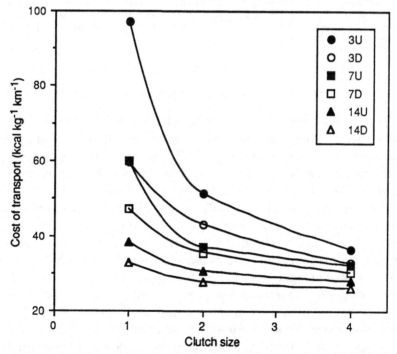

Figure 6. Comparison of costs of transport for clutches of different numbers of individuals. Values were based on mean mass-specific metabolic rates. The insert shows symbols for combinations of the age in days of the ducklings (3, 7, 14) and the position of the decoy (U, up above the water; D, down in the water).

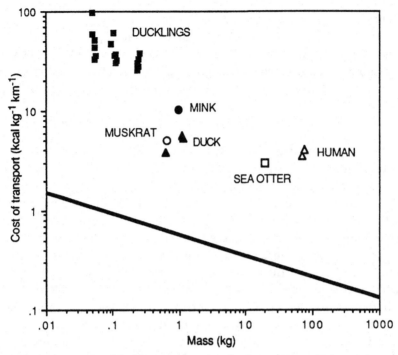

Figure 7. Cost of transport of ducklings compared with other swimmers. Other surface paddlers include duck (Prange & Schmidt-Nielsen, 1972; Woakes & Butler, 1983; Baudinette & Gill, 1985), human (Di Pramero, personal communication, 1979), mink (Williams, 1983), muskrat (Fish, 1982), and sea otter (Williams, 1989). The solid line is the minimum cost of transport extrapolated from data on fish (Davis *et al.*, 1985).

DISCUSSION

Although ducklings experience increased energy economy when swimming in formation, it should be cautioned that other reasons may have led ultimately to this behaviour. Formation movement by various animal species has been related to protection against predation, locating food resources, mating efficiency, pooling orientation information, greater tolerance to toxic substances and increasing energy economy (Brock & Riffenburgh, 1960; Breder, 1967, 1976; Weihs, 1973; Shaw, 1978; Hummel, 1983). Formation movement by ducklings is advantageous where swimming performance is potentially limited by energy availability. A reduction in energy expenditure and drag during formation swimming would allow ducklings as a group to swim for longer durations and at higher speeds than as individuals.

Limitations on performance on ducklings may be severe, because of their swimming mode. Surface swimming by ducklings is accomplished by using alternate strokes of the webbed hind feet in a paddle-propulsive mode. Although repetitive paddling is used at slow speeds and for surface swimming, this mode is uneconomical due to its high drag and corresponding inefficiency (Prange & Schmidt-Nielsen, 1970; Fish, 1982, 1984b, 1992; Baudinette & Gill, 1985; Stephenson *et al.*, 1989).

Swimming at the water surface can limit performance by increasing the drag experienced by the body due to an additional wave drag component. Wave drag can increase total drag by a factor of five (Hertel, 1966). In addition, maximum cruising speed is limited by interaction of waves created by the body. The maximum sustainable speed of mallard ducks (Prange & Schmidt-Nielsen, 1970) and muskrats (Fish, 1982) coincided with their predicted hull speeds based on the performance of surface craft. Hull speed is the practical limit to the speed of a displacement hull at the water surface due to the increased drag from wave formation. Hull speed is dependent on the waterline length of the body, so that hull speeds for small bodies are lower than for large bodies. The configuration of duckling formations may effectively increase the waterline length for the entire clutch and allow ducklings to achieve higher sustained swimming speeds as a group than individually.

The reduced energy economy associated with increasing age of the ducklings indicates the importance of inertial effects in formation swimming. When dye was injected behind the decoy, the rotation pattern of the vortices was similar to the condition illustrated in Figure 1B. The momentum transported in the water by the vorticity could be imparted to the trailing ducklings. Because momentum is directly proportional to mass, for a given amount of momentum transferred a smaller duckling will be more effectively pulled along behind the decoy than a larger duckling. This was observed particularly when comparing the metabolic effect on single ducklings swimming with or without the wake of the decoy. The energy saving by 3-d-old ducklings in the decoy's wake was over twice as great as for a 14-d-old duckling which was 4·5 times the mass. A similar situation is observed in porpoises where a smaller animal, often an infant, positions itself close to and slightly behind a larger porpoise to obtain a free ride (Norris & Prescott, 1961). In motor pacing, a human cyclist drafts behind a more massive body such as an automobile and achieves speeds over 220 km h^{-1} (Gross et al., 1983).

In addition to the beneficial vortex pattern established in the drag wake of the decoy, the thrust wake produced by the paddling movements of the ducklings feet may aid in reducing swimming costs. As a foot is swept posteriorly during the power phase of the paddling stroke, a pair of vortices will be generated with one vortex located directly above the other (Alexander, 1982). The rotation and position of the upper vortex is such that it could transfer momentum and reduce the relative velocity of a trailing duckling. This effect could be even larger for ducklings swimming in the thrust wake of an actively paddling adult.

Energy saved during formation swimming could maximize the net rate of energy gain by making more energy potentially available for growth and other functions. Mallard ducklings experience their highest mortality within 18 d of hatching (Orthmeyer & Ball, 1990). Attainment of adult body size through an increased growth rate benefits individuals by decreasing predation risk, increasing thermoregulatory control, and increasing locomotor performance. The latter point is especially critical for migrating birds that must reach full flight capability in time for autumn departure.

In conclusion, formation movement by ducklings reduces metabolic effort during swimming. The influence of following in the wake of a decoy significantly decreased metabolic swimming effort by ducklings, and was most pronounced for the youngest ducklings and small clutches. These findings suggest that the flow pattern generated by the formation reduces the drag encountered when swimming and allows for conservation of energy on an individual basis.

I am indebted to B.D. Clark, J.M.V. Rayner, M.J. Sheehan and H.M. Tiebout III for their comments on earlier versions of this manuscript, to G.W. Fairchild for statistical assistance and to T.L. Aigeldinger and J.M. Battle for their assistance. The research was supported by a grant from the National Science Foundation (DCB-9117274), and by a Research and Publication Committee Grant, and a College of Arts and Sciences Support and Development Award, West Chester University.

Bibliography

Abel, D.C., Graham, J.B., Lowell, W.R. & Shabetai, R., 1986. Elasmobranch pericardial function. 1. Pericardial pressures are not always negative. *Fish Biochemistry and Physiology*, **1**, 75-83.

Abrahams, M.V. & Colgan, P.W., 1985. Risk of predation, hydrodynamic efficiency and their influence on school structure. *Environmental Biology of Fishes*, **13**, 195-202.

Alexander, R.McN., 1965. The lift produced by the heterocercal tails of Selachii. *Journal of Experimental Biology*, **43**, 131-138.

Alexander, R.McN., 1966. Physical aspects of swimbladder function. *Biological Reviews*, **41**, 141-176.

Alexander, R.McN., 1967. *Functional design in fishes*. London: Hutchinson. [Hutchinson University Library Series.]

Alexander, R.McN., 1968. *Animal mechanics*. Seattle: University of Washington.

Alexander, R.McN., 1969. The orientation of muscle fibres in the myomeres of fishes. *Journal of the Marine Biology Association of the United Kingdom*, **49**, 263-290.

Alexander, R.McN., 1972. The energetics of vertical migration by fishes. *Symposia of the Society for Experimental Biology. Cambridge*, **26**, 273-294.

Alexander, R.McN., 1977. Swimming. In *Mechanics and energetics of animal locomotion* (ed. R.McN. Alexander and G. Goldspink), pp. 222-248. London: Chapman and Hall.

Alexander, R.McN., 1982. *Locomotion of animals*. Glasgow: Blackie.

Alexander, R.McN., 1983. *Animal mechanics*, 2nd edition. Oxford: Blackwell Scientific.

Alexander, R.McN., 1984. The gaits of bipedal and quadrupedal animals. *International Journal of Robotics Research*, **3**, 49-59.

Alexander, R.McN., 1989a. *Dynamics of dinosaurs and other extinct giants*. New York: Columbia University Press.

Alexander, R.McN., 1989b. Optimization and gaits in the locomotion of vertebrates. *Physiological Reviews*, **69**, 1199-1227.

Alexander, R.McN., 1990. Size, speed and buoyancy adaptations in aquatic animals. *American Zoologist*, **30**, 189-196.

Alexander, R. McN., 1991. Apparent adaptation and actual performance. In *Evolutionary biology*, vol. 25 (ed. M.K. Hecht, B. Wallace, and R.J. Macintyre), pp. 357-363. New York: Plenum.

Aleyev, Yu. G., 1977. *Nekton*. The Hague: Junk.

Altringham, J.D. & Johnston, I.A., 1990a. Modelling muscle power output in a swimming fish. *Journal of Experimental Biology*, **148**, 395-402.

Altringham, J.D. & Johnston, I.A., 1990b. Scaling effects on muscle function: power output of isolated fish muscle fibres performing oscillatory work. *Journal of Experimental Biology*, **151**, 453-467.

Altringham, J.D., Wardle, C.S. & Smith, C.I., 1993. Myotomal muscle function at different locations in the body of a swimming fish. *Journal of Experimental Biology*, **182**, 191-206.

Amos, W.B., 1975. Contraction and calcium-binding in vorticellid ciliates. In *Molecule and cell development* (ed. R.E. Stephens and S. Inoue), pp. 411-436. New York: Raven Press.

Amos, L.A. & Amos, W.B., 1991. Molecules of the cytoskeleton. In *Molecular cell biology* (ed. J. Skidmore), pp. 42-207. New York: Guilford Press.

Maddock, L., Bone, Q. & Rayner, J.M.V. (ed.). *Mechanics and Physiology of Animal Swimming*.
© 1994. Cambridge University Press.

Amos, W.B., Routledge, L.M., Weis Vogh, T. & Yew, F.F., 1976. The spasmoneme and calcium-dependent contraction in connection with specific calcium binding proteins. In *Calcium in biological systems*, vol. 3 (ed. C.J. Duncan), pp. 273-301. Cambridge University Press.

Anderson, M.E. & Johnston, I.A., 1992. Scaling of power output in fast muscle fibres of the Atlantic cod during cyclical contractions. *Journal of Experimental Biology*, **170**, 143-154.

Anderson, O.R., 1980. Biochemistry and physiology of Protozoa. In *Radiolaria* (ed. M. Levandowsky and S.H. Hutner), pp. 1-40. New York: Academic Press.

Anderson, O.R., 1983. *Radiolaria*. New York: Springer Verlag.

Anderson, O.R., 1987. *Comparative protozoology*. Berlin: Springer Verlag.

Andrews, C.W., 1910-1913. *A descriptive catalogue of the marine reptiles of the Oxford clay*, 2 volumes. London: British Museum (Natural History).

Appleby, R.M., 1979. The affinities of Liassic and later ichthyosaurs. *Palaeontology*, **22**, 921-946.

Archer, S.D., Altringham, J.D. & Johnston, I.A., 1990. Scaling effects on the neuromuscular system, twitch kinetics and morphometrics of the cod, *Gadus morhua*. *Marine Behaviour and Physiology*, **17**, 137-146.

Archer, S.D. & Johnston, I.A., 1989. Kinematics of labriform and subcarangiform swimming in the Antarctic fish *Notothenia neglecta*. *Journal of Experimental Biology*, **143**, 195-210.

Arnold, G.P., 1969. The reactions of plaice (*Pleuronectes platessa*) to water currents. *Journal of Experimental Biology*, **51**, 681-697.

Arnold, G.P., Webb, P.W. & Holford, B.H., 1991. The role of the pectoral fins in station-holding of Atlantic salmon parr (*Salmo salar* L). *Journal of Experimental Biology*, **156**, 625-629.

Arnold, G.P. & Weihs, D., 1978. The hydrodynamics of rheotaxis in the plaice (*Pleuronectes platessa* L.). *Journal of Experimental Biology*, **75**, 147-169.

Au, D. & Weihs, D., 1980. At high speeds dolphins save energy by leaping. *Nature, London*, **284**, 548-550.

Badgerow, J.P. & Hainsworth, F.R., 1981. Energy savings through formation flight? A re-examination of the vee formation. *Journal of Theoretical Biology*, **93**, 41-52.

Bainbridge, R., 1958. The speed of swimming of fish as related to size and to frequency of the tail beat. *Journal of Experimental Biology*, **35**, 109-133.

Bainbridge, R., 1960. Speed and stamina in three fish. *Journal of Experimental Biology*, **37**, 129-153.

Bainbridge, R., 1961. Problems of fish locomotion. *Symposia of the Zoological Society of London*, **40**, 23-56.

Bakker, R.T., 1975. Experimental and fossil evidence for the evolution of tetrapod energetics. In *Ecological studies, analysis and synthesis*, vol. 12 (ed. D.M. Gates and R.B. Schmerl), pp. 365-399. New York: Springer Verlag.

Baldwin, J., 1988. Predicting the swimming and diving behaviour of penguins from muscle biochemistry. *Hydrobiologia*, **165**, 255-261.

Bannasch, R., 1986a. Morphologisch-funktionelle Untersuchungen am Lokomotionsapparat der Pinguine als Grundlage für ein allgemeines Modell des 'Unterwasserfluges', Teil I u. II. *Gegenbaurs Morphologisches Jahrbuch. Leipzig*, **132**, 645-679.

Bannasch, R., 1986b. Morphologisch-funktionelle Untersuchungen am Lokomotionsapparat der Pinguine als Grundlage für ein allgemeines Modell des 'Unterwasserfluges', Teil I u. II. *Gegenbaurs Morphologisches Jahrbuch. Leipzig*, **132**, 757-817.

Bannasch, R., 1987. Morphologisch-funktionelle Untersuchungen am Lokomotionsapparat der Pinguine als Grundlage für ein allgemeines Modell des 'Unterwasserfluges', Teil III. *Gegenbaurs Morphologisches Jahrbuch. Leipzig*, **132**, 39-59.

Bannasch, R., in press. Hydrodynamics on penguins - an experimental approach. In *Penguins* (ed. P. Dann *et al.*). Sydney: Surrey Beatty. [Proceedings of the Second International Penguin Conference.]

Bannasch, R. & Fiebig, J., 1992. Herstellung von Pinguinmodellen für hydrodynamische Untersuchungen. *Der Präparator, Bochum*, **38**, 1-5.

Batchelor, G.K., 1967. *An introduction to fluid dynamics*. Cambridge University Press.

Batty, R.S., 1981. Locomotion of plaice larvae. *Symposia of the Zoological Society of London*, **5**, 13-32.

Batty, R.S., 1984. Development of swimming movements and musculature of larval herring (*Clupea harengus*). *Journal of Experimental Biology*, **110**, 217-229.

Baudinette, R.V. & Gill, P., 1985. The energetics of 'flying' and 'paddling' in water: locomotion in penguins and ducks. *Journal of Comparative Physiology*, **155B**, 373-380.

Baumel, J.J., King, A.S., Lucas, A.M., Breazile, J.E. & Evans, H.E. (ed.), 1979. *Nomina anatomica avium*. Academic Press.

Bé, A.W.H. & Anderson, O.R., 1976. Gametogenesis in planktonic Foraminifera. *Science, New York*, **192**, 890-892.

Beamish, F.W.H., 1978. Swimming capacity. In *Fish physiology*. Vol. 7. *Locomotion* (ed. W.S. Hoar and D. J. Randall), pp. 101-187. New York: Academic Press.

Bellrose, F.C., 1976. *Ducks, geese and swans of North America*. Harrisburg, Pennsylvania: Stackpole.

Belyayev, V.V. & Zuyev, G.V., 1969. Hydrodynamic hypothesis of school formation in fishes. *Problems of Ichthyology*, **9**, 578-584.

Bennett, A.F., 1985. Energetics and locomotion. In *Functional vertebrate morphology* (ed. M. Hildebrand et al.), pp. 173-184. Cambridge, Mass.: Harvard University Press.

Bennett, A.F., 1991. The evolution of activity capacity. *Journal of Experimental Biology*, **160**, 1-23.

Bennett, M.B., 1992. Empirical studies of walking and running. *Advances in Comparative Environmental Physiology*, **11**, 141-165.

Benton, M.J., 1990. *Vertebrate paleontology*, pp. 113-117, 194-199. London: Unwin Hyman.

Berg, H.C., 1983. *Random walks in biology*. Princeton University Press.

Bill, R.G. & Herrnkind, W.F. 1976. Drag reduction by formation movement in spiny lobsters. *Science, New York*, **193**, 1146-1149.

Blake, R.W., 1976. On seahorse locomotion. *Journal of the Marine Biological Association of the United Kingdom*, **56**, 939-949.

Blake, R.W., 1977. On ostraciiform locomotion. *Journal of the Marine Biological Association of the United Kingdom*, **57**, 1047-1055.

Blake, R.W., 1978. On balistiform locomotion. *Journal of the Marine Biological Association of the United Kingdom*, **58**, 73-80.

Blake, R.W., 1979. The energetics of hovering in the mandarin fish (*Synchropus picturatus*). *Journal of Experimental Biology*, **82**, 25-33.

Blake, R.W., 1980. The mechanics of labriform locomotion. II. An analysis of the recovery stroke and the overall fin-beat cycle propulsive efficiency in the angelfish. *Journal of Experimental Biology*, **85**, 337-342.

Blake, R.W., 1983a. *Fish locomotion*. Cambridge University Press.

Blake, R.W., 1983b. Swimming in the electric eels and knifefishes. *Canadian Journal of Zoology*, **61**, 1432-1441.

Blake, R.W., 1983c. Energetics of leaping in dolphins and other aquatic animals. *Journal of the Marine Biological Association of the United Kingdom*, **63**, 61-70.

Blake, R.W., 1983d. Median and paired fin propulsion. In *Fish biomechanics* (ed. P.W. Webb and D. Weihs), pp. 214-247. New York: Praeger.

Blake, R.W. & Smith, M.D., 1988. On penguin porpoising. *Canadian Journal of Zoology*, **66**, 2093-2094.

Blight, A.R., 1977. The muscular control of vertebrate swimming movements. *Biological Reviews*, **52**, 181-218.

Blight, A.R., 1976. Undulating swimming with or without waves of contraction. *Nature, London*, **264**, 352-354.

Block, B.A., Booth, D. & Carey, F.G., 1992. Direct measurements of swimming speeds and depth of blue marlin. *Journal of Experimental Biology*, **166**, 267-284.

Block, B.A., Finnerty, J.R., Stewart, A.F.R. & Kidd, J., 1993. Evolution of endothermy in fish: mapping physiological traits on a molecular phylogeny. *Science, New York*, **260**, 210-214.

Boggs, C.H. & Kitchell, J.F., 1991. Tuna metabolic rates estimated from energy losses during starvation. *Physiological Zoology*, **64**, 502-524.

Boisclair, D. & Tang, M., 1993. Empirical analysis of the swimming pattern on the net energetic cost of swimming in fishes. *Journal of Fish Biology*, **42**, 169-183.

Boletzky, S. von, 1987. Juvenile behaviour. In *Cephalopod life cycles*, vol. 2 (ed. P.R. Boyle), pp. 45-60. London: Academic Press.

Bolhuis, J.J., 1991. Mechanisms of avian imprinting: a review. *Biological Reviews of the Cambridge Philosophical Society*, **66**, 303-345.

Bone, Q., 1966. On the function of the two types of myotomal muscle fibre in elasmobranch fish. *Journal of the Marine Biological Association of the United Kingdom*, **46**, 321-349.

Bone, Q., 1978. Locomotor muscle. In *Fish physiology*. Vol. 7. *Locomotion* (ed. W.S. Hoar and D.J. Randall), pp. 361-424. New York: Academic Press.

Bone, Q., 1988. Muscles and locomotion. In *Physiology of elasmobranch fishes* (ed. T.J. Shuttleworth), pp. 99-141. Berlin: Springer-Verlag.

Bone, Q., Kecenuik, J. & Jones, D.R., 1978. On the role of different fibre types in fish myotomes at intermediate swimming speeds. *Fishery Bulletin. National Oceanic and Atmospheric Administration. Washington, DC*, **76**, 691-699.

Bone, Q., Kiceniuk, J. & Jones, D.R., 1978. On the role of the different fibre types in fish myotomes at intermediate swimming speeds. *Fishery Bulletin. National Oceanic and Atmospheric Administration. Washington DC*, **76**, 691-699.

Bone, Q. & Marshall, N.B., 1982. *Biology of fishes*. London: Blackie.

Bone, Q., Pulsford, A. & Chubb, A.D., 1981. Squid mantle muscle. *Journal of the Marine Biological Association of the United Kingdom*, **61**, 327-342.

Bowtell, G. & Williams, T.L., 1991. Anguilliform body dynamics: modelling the interaction between muscle activation and body curvature. *Philosophical Transactions of the Royal Society of London* (B), **334**, 385-390.

Bowtell, G. & Williams, T.L., 1993. Anguilliform body dynamics: a continuum model for the interaction between muscle activation and body curvature. *Journal of Mathematical Biology*, in press.

Boycott, B.B., 1961. The functional organization of the brain of the cuttlefish *Sepia officinalis*. *Proceedings of the Royal Society of London* (B), **153**, 503-534.

Bradbury, H.E. & Aldrich, F.A., 1969. Observations on locomotion of the short-finned squid, *Illex illecebrosus illecebrosus* (Lesueur, 1821), in captivity. *Canadian Journal of Zoology*, **47**, 741-744.

Brandt, I.F., 1840. Beiträge zur Kenntnis der Naturgeschichte der Vögel. 5. Abhandlung. Über die Flossentaucher (impennes seu Aptenodytidae) als Typen einer Gruppe unter den Schwimmvögeln. *Mémoires de l'Académie Impériale des Sciences de St Petersbourg*, 6e série, **3**, 213-217.

Braun, J. & Reif, W.-E., 1982. A new terminology of aquatic propulsion in vertebrates. *Neues Jahrbuch für Geologie und Paläontologie, Abhandlungen*, **164**, 162-167.

Braun, J. & Reif, W., 1985. A survey of aquatic locomotion in fishes and tetrapods. *Neues Jahrbuch für Geologie und Paläontologie. Stuttgart. Abhandlungen*, **169**, 307-332.

Breder, C.M. Jr, 1926. The locomotion of fishes. *Zoologica*, **4**, 159-297.

Breder, C.M. Jr, 1965. Vortices and fish schools. *Zoologica*, **50**, 97-114.

Breder, C.M. Jr, 1967. On the survival value of fish schools. *Zoologica*, **52**, 25-40.

Breder, C.M. Jr, 1976. Fish schools as operational structures. *Fishery Bulletin. National Oceanic and Atmospheric Administration. Washington DC*, **74**, 471-502.

Brett, J.R., 1965. The relation of size to rate of oxygen consumption and sustained swimming speed of sockeye salmon (*Oncorhynchus nerka*). *Journal of the Fisheries Research Board of Canada*, **22**, 1491-1501.

Brett, J.R., 1979. Environmental factors and growth. In *Fish physiology*. Vol. 8. *Bioenergetics and growth* (ed. W.S. Hoar *et al.*), pp. 599-675. New York: Academic Press.

Brett, J.R. & Blackburn, J.M., 1978. Metabolic rate and energy expenditure of the spiny dogfish, *Squalus acanthias*. *Journal of the Fisheries Research Board of Canada*, **35**, 816-821.

Brett, J.R. & Groves, T.D.D., 1979. Physiological energetics. In *Fish physiology*. Vol. 8. *Bioenergetics and growth* (ed. W.S. Hoar *et al.*), pp. 279-352. New York: Academic Press.

Brill, R.W., 1987. On the standard metabolic rates of tropical tunas, including the effect of body size and acute temperature change. *Fishery Bulletin. National Oceanic and Atmospheric Administration. Washington, DC*, **85**, 25-36.

Brill, R.W. & Bushnell, P.G., 1991. Metabolic and cardiac scope of high energy demand teleosts, the tunas. *Canadian Journal of Zoology*, **69**, 2002-2009.

Brill, R.W. & Dizon, A.E., 1979. Red and white muscle fibre activity in swimming skipjack tuna, *Katsuwonus pelamis* (L.). *Journal of Fish Biology*, **15**, 679-685.

Brock, V.E. & Riffenburgh, R.H., 1960. Fish schooling: a possible factor in reducing predation. *Journal du Conseil Permanent International pour l'Exploration de la Mer*, **25**, 307-317.

Brown, D.S., 1981. The English Upper Jurassic Plesiosauroidea (Reptilia), and a review of the phylogeny and classification of the Plesiosauria. *Bulletin of the British Museum (Natural History), Geology*, **35**, 253-347.

Buchanan, J.T. & Grillner, S., 1987. Newly identified 'glutamate interneurons' and their role in locomotion in the lamprey spinal cord. *Science, New York*, **236**, 312-314.

Buffrénil, V. de, Collet, A. & Pascal, M., 1985. Ontogenetic development of skeletal weight in a small delphinid, *Delphinus delphis* (Cetacea, Odontoceti). *Zoomorphology*, **105**, 336-344.

Buffrénil, V. de & Mazin, J.-M., 1989. Bone histology of *Claudiosaurus germaini* (Reptilia, Claudiosauridae) and the problem of pachyostosis in aquatic tetrapods. *Historical Biology*, **2**, 311-322.

Buffrénil, V. de & Mazin, J.-M., 1990. Bone histology of the ichthyosaurs: comparative data and functional interpretation. *Paleobiology*, **16**, 435-447.

Buffrénil, V. de & Schoevaert, D., 1989. Données quantitatives et observations histologiques sur la pachyostose du squelette du dugong (*Dugong dugon*) (Müller) (Sirenia, Dugongidae). *Canadian Journal of Zoology*, **67**, 2107-2119.

Buffrénil, V. de, Sire, J.-Y. & Schoevaert, D., 1986. Comparaison de la structure et du volume squelettiques entre un delphinidé (*Delphinus delphis* L.) et un mammifère terrestre (*Panthera leo* L.). *Canadian Journal of Zoology*, **64**, 1750-1756.

Bushnell, P.G. & Brill, R.W., 1992. Oxygen transport and cardiovascular responses in skipjack tuna (*Katsuwonus pelamis*) and yellowfin tuna (*Thunnus albacares*) exposed to acute hypoxia. *Journal of Comparative Physiology B*, **162**, 131-143.

Bushnell, P.G., Lutz, P.L., Steffensen, J.F., Oikari, A. & Gruber, S.H., 1982. Increases in arterial blood oxygen during exercise in the lemon shark (*Negaprion brevirostris*). *Journal of Comparative Physiology A*, **147**, 41-47.

Bushnell, P.G., Steffensen, J.F. & Johansen, K., 1984. Oxygen consumption and swimming performance in hypoxia-acclimated rainbow trout, *Salmo gairdneri*. *Journal of Experimental Biology*, **113**, 225-235.

Cachon, J. & Cachon, M., 1984. Adaptative evolution of protists to planktonic life. *Annales de l'Institut Océanographique*, **60**, 105-114.

Cachon, J. & Cachon, M., 1985. Non-actin filaments and cell contraction in *Kofoidinium* and other dinoflagellates. *Cell Motility*, **5**, 1-15.

Cachon, J. & Cachon, M., 1986. Adaptation des dinoflagellés à la vie planctonique. *Bolletino di Zoologia, Unione Zoologica Italiano*, **53**, 239-245.

Cachon, J., Cachon, M. & Boillot, A., 1983. Flagellar rootlets as myonemal elements for pusule contractility in dinoflagellates. *Cell Motility*, **3**, 61-77.

Cachon, M., Cachon, J., Cosson, J., Greuet, C. & Huitorel, P., 1991. Dinoflagellate flagella adopt various conformations in response to different needs. *Biologie Cellulaire*, **71**, 175-182.

Cachon, J., Cachon, M. & Greuet, C., 1970. Le système pusulaire de quelques péridiniens libres ou parasites. *Protistologica*, **6**, 467-476.

Cachon, J., Cachon, M., Boillot, A. & Brown, D.L., 1987. Cytoskeletal and myonemal structures of dinoflagellates are made of 2-3 nm filaments. *Cell motility and the cytoskeleton*, **7**, 325-336.

Cachon, J., Cachon, M., Tilney, L.G. & Tilney, M.S., 1977. Movement generated by interactions between the dense material at the ends of microtubules and non-actin-containing microfilaments in *Sticholonche zanclea*. *Journal of Cell Biology*, **72**, 314-338.

Cachon, M., Cosson, J., Cosson, M-P., Huitorel, P. & Cachon, J., 1988. Ultrastructure of the flagellar apparatus of *Oxyrrhis marina*. *Biologie Cellulaire*, **63**, 159-168.

Caldwell, D.K. & Caldwell, M.C., 1985. Manatees *Trichechus manatus* Linnaeus, 1758; *Trichechus senegalensis* Link, 1795 and *Trichechus inunguis* (Natterer, 1883). In *Handbook of marine mammals.* Vol. 3. *The sirenians and baleen whales* (ed. S.H. Ridgway and R.J. Harrison), pp. 33-66. London: Academic Press.

Camp, C.L., 1980. Large ichthyosaurs from the Upper Triassic of Nevada. *Palaeontographica* A, **170**, 139-200.

Carling, J.C. & Williams, T.L., 1991. Lamprey hydrodynamics: the numerical solution of the Navier-Stokes equations. (Abstract.) *Journal of the Marine Biological Association of the United Kingdom*, **71**, 711.

Carroll, R.L., 1988. *Vertebrate paleontology and evolution.* New York: Freeman.

Casey, T.M., 1992. Energetics of locomotion. *Advances in Comparative Environmental Physiology*, **11**, 251-275.

Cater, B., 1989. Deep in the forest. *BBC Wildlife*, **12**, 792-804.

Cavagna, G.A., Mazzanti, M., Heglund, N.C. & Citterio, G., 1985. Storage and release of mechanical energy by active muscle: a non-elastic mechanism. *Journal of Experimental Biology*, **115**, 79-87.

Childress, S., 1981. *Mechanics of swimming and flying.* Cambridge University Press.

Clark, B.D. & Bemis, W., 1979. Kinematics of swimming of penguins at the Detroit Zoo. *Journal of Zoology*, **188**, 411-428.

Clarke, M.R., 1966. A review of the systematics and ecology of oceanic squids. *Advances in Marine Biology*, **4**, 91-300.

Clarke, M.R., 1978a. Structure and proportions of the spermaceti organ in the sperm whale. *Journal of the Marine Biological Association of the United Kingdom*, **58**, 1-17.

Clarke, M.R., 1978b. Physical properties of spermaceti oil in the sperm whale. *Journal of the Marine Biological Association of the United Kingdom*, **58**, 19-26.

Clarke, M.R., 1978c. Buoyancy control as a function of the spermaceti organ in the sperm whale. *Journal of the Marine Biological Association of the United Kingdom*, **58**, 27-71.

Clarke, M.R., 1988. Evolution of buoyancy and locomotion in recent cephalopods. In *The Mollusca*, vol. 12 (ed. M.R. Clarke and E.R. Trueman), pp. 203-213. San Diego: Academic Press.

Clarke, M.R., Denton, E.J. & Gilpin-Brown, J.B., 1979. On the use of ammonium for buoyancy in squids. *Journal of the Marine Biological Association of the United Kingdom*, **59**, 259-276.

Cockcroft, V.G. & Ross, G.J.B., 1990. Observations on the early development of a captive bottlenose dolphin calf. In *The bottlenose dolphin* (ed. S. Leatherwood and R.R. Reeves), pp. 461-478. San Diego: Academic Press.

Cohen, A.H., Ermentrout, B.E., Kiemel, T., Kopell, N., Mellen, N., Sigvardt, K.A. & Williams, T.L., 1992. Modelling of intersegmental co-ordination in the lamprey central pattern generator for locomotion. *Trends in Neurosciences*, **15**, 434-438.

Cole, K.S. & Gilbert, D.L., 1970. Jet propulsion of squid. *Biological Bulletin. Marine Biological Laboratory, Woods Hole*, **138**, 245-246.

Compagno, L.J.V., 1977. Phyletic relationships of living sharks and rays. *American Zoologist*, **17**, 303-322.

Compagno, L.J.V., 1988. *Sharks of the order Carcharhiniformes.* Princeton University Press.

Cone, C.D. Jr, 1962. Thermal soaring of birds. *American Scientist*, **50**, 180-209.

Cosson, J., Cachon, M., Cachon, J. & Cosson, M-P., 1988. Swimming behaviour of the unicellular biflagellate *Oxyrrhis marina*: in vivo and in vitro movement of the two flagella. *Biologie Cellulaire*, **63**, 117-126.

Cott, H.B., 1961. Scientific results of an inquiry into the ecology and economic status of the Nile crocodile (*Crocodilus niloticus*) in Uganda and Northern Rhodesia. *Transactions of the Zoological Society of London*, **29**, 211-357.

Coues, E., 1872. Material for a monograph of Spheniscidae. *Proceedings of the Academy of Natural Sciences of Philadelphia*, **1872**, 170-212.

Culik, B.M., Wilson, R.P. & Bannasch, R., in press. Under-water swimming at low energetic cost by Pygoscelid penguins. *Journal of Experimental Biology*.

Currey, J.D., 1984. *The mechanical adaptations of bones.* Princeton University Press.

Curtin, N.A. & Woledge, R.C., 1988. Power output and force-velocity relationship of live fibres from white myotomal muscle of the dogfish *Scyliorhinus canicula*. *Journal of Experimental Biology*, **140**, 187-197.

Curtin, N.A. & Woledge, R.C., 1991. Efficiency of energy conversion during shortening of muscle fibres from the dogfish *Scyliorhinus canicula*. *Journal of Experimental Biology*, **158**, 343-353.

Cuvier, G., 1800. *Leçons díanatomie comparée*. Volume 1. Paris.

Daniel, T.L., 1983. Mechanics and energetics of medusan jet propulsion. *Canadian Journal of Zoology*, **61**, 1406-1420.

Daniel, T.L., 1984. Unsteady aspects of aquatic locomotion. *American Zoologist*, **24**, 121-134.

Daniel, T.L., 1985. Cost of locomotion: unsteady medusan swimming. *Journal of Experimental Biology*, **119**, 149-164.

Daniel, T.L., 1988. Forward flapping flight from flexible fins. *Canadian Journal of Zoology*, **66**, 630-638.

Daniel, T.L., Jordan, C. & Grunbaum, B., 1992. Hydromechanics of swimming. In *Advances in comparative and environmental physiology*. Vol 11. *Mechanics of animal locomotion* (ed. R.McN. Alexander), pp. 17-49. Berlin: Springer-Verlag.

Daniel, T.L. & Webb, P.W., 1987. Physics, design and locomotor performance. In *Comparative physiology: life in water and on land* (ed. P. Dejours et al.), pp. 343-369. New York: Liviana Press, Springer-Verlag.

Davenport, J., Grove, D.J., Cannon, J., Ellis, T.R. & Stables, R., 1990. Food capture, appetite, digestion rate and efficiency in hatchling and juvenile *Crocodylus porosus*. *Journal of Zoology*, **220**, 569-592.

Davenport, J., Munks, S.A. & Oxford, P.J., 1984. A comparison of the swimming of marine and fresh-water turtles. *Proceedings of the Royal Society of London* B, **220**, 447-475.

Davidson, L.A., 1982. Ultrastructure, behavior and algal flagellate affinities of the helioflagellate *Ciliophrys marina* and the classification of the helioflagellates (Protista, Actinopoda, Heliozoea). *Journal of Protozoology*, **29**, 19-29.

Davies, M. & Johnston, I.A., 1993. Muscle fibres in rostral and caudal myotomes of the Atlantic cod have different contractile properties. *Journal of Physiology*, **459**, 8P.

Davis, R.W., Williams, T.M. & Kooyman, G.L., 1985. Swimming metabolism of yearling and adult harbor seals *Phoca vitulina*. *Physiological Zoology*, **58**, 590-596.

Davison, W. & Macdonald, J.A., 1985. A histochemical study of the swimming musculature of Antarctic fish. *New Zealand Journal of Zoology*, **12**, 473-483.

Deitmer, J.W., 1989. Ion channels and the cellular behavior of *Stylonychia*. In *Evolution of the first nervous systems* (ed P. Anderson), pp. 255-265. New York: Plenum Press. [Life Sciences, 188.]

Dentler, W.L., 1987. Cilia and flagella. *International Review of Cytology*, supplement 17, 91-456.

Denton, E.J., 1974. On buoyancy and the lives of modern and fossil cephalopods. *Proceedings of the Royal Society of London* (B), **185**, 273-299.

Dewar, H., 1992. The swimming energetics of yellowfin tuna. *American Zoologist*, **32**, 34A.

Dewar, H. & Graham, J.B., in press a. Studies of tropical tuna swimming performance in a large water tunnel. I: Energetics. *Journal of Experimental Biology*.

Dewar, H. & Graham, J.B., in press b. Studies of tropical tuna swimming performance in a large water tunnel. III: Kinematics. *Journal of Experimental Biology*.

Dewar, H., Graham, J.B. & Brill, R.W., 1991. Physiological thermoregulation in the yellowfin tuna, *Thunnus albacares*. *American Zoologist*, **31**, 41A.

Dewar, H., Graham, J.B. & Brill, R.W., in press. Studies of tropical tuna swimming performance in a large water tunnel. II: Thermoregulation. *Journal of Experimental Biology*.

Dinnendahl, L. & Kramer, G., 1957. Über größenabhängige Änderungen von Körperproportionen bei Möwen (*Larus ridibundus, L. canus, L. argentatus, L. marinus*). *Journal für Ornithologie.*, **98**, 282-312.

Dizon, A.E. & Brill, R.W., 1979. Thermoregulation in tunas. *American Zoologist*, **19**, 249-265.

Dodge, J.D. & Greuet, C., 1987. Dinoflagellate ultrastructure and complex organelles. In *The biology of dinoflagellates* (ed. F.R.J. Taylor), pp. 92-142. Oxford: Blackwells. [Botanical Monographs, vol. 21.]

Domenici, P. & Blake, R.W., 1991. The kinematics and performance of the escape in the angelfish (*Pterophyllum eimekei*). *Journal of Experimental Biology*, **156**, 187-205.

Domning, D.P. & Buffrénil, V. de, 1991. Hydrostasis in the Sirenia: quantitative data and functional interpretations. *Marine Mammal Science*, **7**, 331-368.

Dunlap, K., 1977. Localization of calcium channels in *Paramecium caudatum*. *Journal of Physiology*, **271**, 119-133.

Dunson, W.A., 1975. Adaptations of sea snakes. In *The biology of sea snakes* (ed. W.A. Dunson), pp. 3-19. Baltimore: University Park Press.

Dyer, A.B. & Gottlieb, G., 1990. Auditory basis of maternal attachment in ducklings (*Anas platyrhynchos*) under simulated naturalistic imprinting conditions. *Journal of Comparative Psychology*, **104**, 190-194.

Eaton, R.C., Bombardieri, R.A. & Meyer, D.L., 1977. The Mauthner-initiated startle response in teleost fish. *Journal of Experimental Biology*, **66**, 65-81.

Eaton, R.C. & Hackett, J.T., 1984. The role of the Mauthner cell in fast-starts involving escape in teleost fishes. In *Neural mechanisms of startle behavior* (ed. R.C. Eaton), pp. 213-266. New York: Plenum Press.

Edman, K.A.P., 1980. Depression of mechanical performance by active shortening during twitch and tetanus of vertebrate muscle fibres. *Acta Physiologica Scandinavica*, **109**, 15-26.

Ehlinger, T.J. & Wilson, D.S., 1988. Complex foraging polymorphism in bluegill sunfish. *Proceedings of the National Academy of Sciences of the United States of America*, **85**, 1878-1882.

Fancy, S.G. & White, R.G., 1985. Incremental cost of activity. In *Bioenergetics of wild herbivores* (ed. R.J. Hudson and R.G. White), pp. 143-159. Boca Raton, Florida: CRC Press.

Farrell, A.P., 1991. From hagfish to tuna - a perspective on cardiac function. *Physiological Zoology*, **64**, 1137-1164.

Farrell, A.P. & Jones, D.R., 1992. The heart. In *Fish physiology*, vol. XII (ed. W.S. Hoar, D.J. Randall and A.P. Farrell), pp. 1-88. New York: Academic Press.

Febvre, J., 1981. The myoneme of the Acantharia (Protozoa), a new model of cellular motility. *Biosystems*, **14**, 327-336.

Febvre, J., 1990. Phylum Actinopoda. Class Acantharia. In *Handbook of Protoctista* (ed L. Margulis *et al.*), pp. 363-379. Boston: Jones & Bartlett.

Febvre, J. & Febvre-Chevalier, C., 1982. Motility processes in Acantharia (Protozoa). I. Cinematographic and cytological study of the myonemes. Evidence for a helix-coil mechanism of the constituent filaments. *Biologie Cellulaire*, **44**, 283-304.

Febvre, J. & Febvre-Chevalier, C., 1989a. Motility processes in Acantharia. II. A Ca^{2+} dependent system of contractile 2-4 nm filaments isolated from demembranated myonemes. *Biologie Cellulaire*, **67**, 243-249.

Febvre, J. & Febvre-Chevalier, C., 1989b. Motility processes in Acantharia. III. Calcium regulation of the contraction-relaxation cycles of *in vivo* myonemes. *Biologie Cellulaire*, **67**, 251-261.

Febvre, J., Febvre-Chevalier, C. & Sato., H., 1990. Polarizing microscope study of a contractile nanofilament system: the acantharian myoneme. *Biologie Cellulaire*, **69**, 41-51.

Febvre-Chevalier, C. & Febvre, J., 1986. Motility mechanisms in actinopods (Protozoa): a review with particular attention to axopodial contraction-extension, and movement of nonactin filament systems. *Cell Motility and the Cytoskeleton*, **6**, 198-208.

Feder, M.E. & Lauder, G.V., 1986. *Predator-prey relationships: perspectives and approaches from the study of lower vertebrates*. Chicago University Press.

Feldkamp, S.D., 1987. Foreflipper propulsion in the California sea lion, *Zalophus californianus*. *Journal of Zoology*, **212**, 43-57.

Fetcho, J.R., 1987. A review of the organization and evolution of motoneurons innervating the axial musculature of vertebrates. *Brain Research Reviews*, **12**, 243-280.

Fields, P.A., 1990. Decreased swimming effort in groups of Pacific mackerel (*Scomber japanicus*). *American Zoologist*, **30**, 134A.

Fierstine, H.L. & Walters, V., 1968. Studies in locomotion and anatomy of scombroid fishes. *Memoirs of the Southern California Academy of Sciences*, **6**, 1-31.

Filhol, M.H., 1882a. Anatomie des manchots. *Bulletin de la Société Philomatique de Paris*, Série 7 (6), 226-248.

Filhol, M.H., 1882b. Anatomie des manchots. *Bulletin de la Société Philomatique de Paris*, Série 7 (7), 16-19, 92-94.

Filhol, M.H., 1884. Anatomie des manchots. *Bulletin de la Société Philomatique de Paris*, Série 7 (8), 60-62.

Filhol, M.H., 1885. Observation relatives à l'anatomie de diverses espèces de manchots. *Recherches zoologiques, botaniques et géologiques faites à l'Ile Campell et en Nouvelle Zélande*. Vol. 3, pp. 65-339. Paris: Gauthier-Villars.

Fish, F.E., 1982. Aerobic energetics of surface swimming in the muskrat *Ondatra zibethicus*. *Physiological Zoology*, **55**, 180-189.

Fish, F.E., 1984a. Kinematics of undulatory swimming in the American alligator. *Copeia*, **1984**, 839-843.

Fish, F.E., 1984b. Mechanics, power output and efficiency of the swimming muskrat (*Ondatra zibethicus*). *Journal of Experimental Biology*, **110**, 183-201.

Fish, F.E., 1987. Kinematics and power output of jet propulsion by the frogfish genus *Antennarius* (Lophiiformes: Antennariidae). *Copeia*, **4**, 1046-1048.

Fish, F.E., 1990. Wing design and scaling of flying fish with regard to flight performance. *Journal of Zoology*, **221**, 391-403.

Fish, F.E., 1992. Aquatic locomotion. In *Mammalian energetics: interdisciplinary views of metabolism and reproduction* (ed. T.E. Tomasi and T.H. Horton), pp. 34-63. Ithaca, New York: Cornell University Press.

Fish, F.E., Fegely, J.F. & Xanthopoulos, C.J., 1991. Burst-and-coast swimming in schooling fish with implications for energy economy. *Comparative Biochemistry and Physiology*, **100**A, 633-637.

Fish, F.E., Innes, S., & Ronald, K., 1988. Kinematics and estimated thrust production of swimming harp and ringed seals. *Journal of Experimental Biology*, **137**, 157-173.

Fish, F.E. & Stein, B.R., 1991. Functional correlates of differences in bone density among terrestrial and aquatic genera in the family Mustelidae (Mammalia). *Zoomorphology*, **1991**, 339-345.

Foyle, T.P. & O'Dor, R.K., 1988. Predatory strategies of squid (*Illex illecebrosus*) attacking small and large fish. *Marine Behavior and Physiology*, **13**, 155-168.

Freadman, M.A., 1979. Role partitioning of swimming musculature of striped bass, *Morone saxatilis* Walbaum and bluefish, *Pomatomus saltatrix* L. *Journal of Fish Biology*, **15**, 417-423.

Frey, E., 1982. Ecology, locomotion and tail muscle anatomy of crocodiles. *Neues Jahrbuch für Geologie und Paläontologie, Abhandlungen*, **164**, 194-199.

Frey, E. & Riess, J., 1982. Considerations concerning plesiosaur locomotion. *Neues Jahrbuch für Geologie und Paläontologie, Abhandlungen*, **164**, 193-194.

Fuiman, L.A., 1986. Burst-swimming performance of larval zebra danios and the effects of diel temperature fluctuations. *Transactions of the American Fisheries Society*, **115**,143-148.

Fuiman, L.A. & Webb, P.W., 1988. Ontogeny of routine swimming activity and performance in zebra danios (Teleostei: Cyprinidae). *Animal Behaviour*, **36**, 250-261.

Full, R.J. & Koehl, M.A.R., 1993. Drag and lift on running insects. *Journal of Experimental Biology*, **176**, 89-101.

Fürbringer, M., 1888. *Untersuchungen zur Morphologie und Systematik der Vögel*. Amsterdam.

Fürbringer, M., 1902. Zur vergleichenden Anatomie des Brustschulterapparates und der Schultermuskeln. V. Teil. Vögel. *Zeitscrift für Naturwissenschaften*, **36** (NF 29), 289-736.

Gadow, H. & Selenka, E., 1893. Vögel. In *Bronns Klassen und Ordnungen des Thierreichs. II. Systematischer Teil*. Leipzig.

Garrod, A.H., 1873. On certain muscles of the thigh of birds and their value in classification. Part I. *Proceedings of the Zoological Society of London*, 626-644.

Gee, J.H., 1983. Ecological implications of buoyancy control in fish. In *Fish biomechanics* (ed. P.W. Webb and D. Weihs), pp. 140-176. New York: Praeger.

George, J.C. & Berger, A.J., 1966. *Avian myology*. Academic Press.

Gervais, P. & Alix, E., 1877. Ostéologie et myologie des manchots ou Sphéniscidés. *Journal de Zoologie. Paris*, **6**, 424-472.

Gibbons, I.R., 1989. Microtubule based motility: an overview of a fast moving field. In *Cell movement*, vol. 1 (ed. V.I.D. Warner *et al.*), pp. 3-22. New York: Allan R. Liss Inc.

Gilly, W.F., Hopkins, B. & Mackie, G.O., 1991. Development of giant motor axons and neural control of escape responses in squid embryos and hatchlings. *Biological Bulletin. Marine Biological Laboratory, Woods Hole*, **180**, 209-220.

Godfrey, S.J., 1984. Plesiosaur subaqueous locomotion: a reappraisal. *Neues Jahrbuch für Geologie und Paläontologie. Stuttgart. Monatshefte*, **11**, 661-672.

Godfrey, S.J., 1985. Additional observations of subaqueous locomotion in the California sea lion (*Zalophus californianus*). *Aquatic Mammals*, **11** (2), 53-57.

Goldspink, G., 1981. The use of muscles during flying, swimming, and running from the point of view of energy saving. *Symposia of the Zoological Society of London*, **48**, 219-238.

Goodenough, V.W., 1989. Cilia, flagella and the basal apparatus. *Current Opinion in Cell Biology*, **1**, 58-62.

Goodenough, V.W. & Heuser, J.E., 1989. Structure of the soluble and *in situ* ciliary dyneins visualized by quick-freeze, deep-etch microscopy. In *Cell movement*, vol. 1, (ed. V.I.D. Warner *et al.*), pp. 121-140. New York: Allan R. Liss Inc.

Goolish, E.M., 1991. Aerobic and anaerobic scaling in fish. *Biological Reviews*, **66**, 33-56.

Gordon, A.M., Huxley, A.F. & Julian, F.J., 1966. The variation in isometric tension with sarcomere length in vertebrate muscle fibres. *Journal of Physiology*, **184**, 170-192.

Gordon, K.R., 1981. Locomotor behaviour of the walrus (*Odobenus*). *Journal of Zoology*, **195**, 349-367.

Gosline, J.M. & Shadwick, R.E., 1983. The role of elastic energy storage mechanisms in swimming: an analysis of mantle elasticity in escape jetting in the squid, *Loligo opalescens*. *Canadian Journal of Zoology*, **61**, 1421-1431.

Gosline, J.M., Steeves, J.D., Harman, A.D. & DeMont, M.E., 1983. Patterns of circular and radial mantle muscle activity in respiration and jetting of the squid *Loligo opalescens*. *Journal of Experimental Biology*, **104**, 97-109.

Gosline, W.A., 1971. *Functional morphology and classification of teleostean fishes*. Honolulu: University of Hawaii Press.

Goulding, M., 1993. Flooded forests of the Amazon. *Scientific American*, **267** (3), 114-120.

Graham, J.B., 1983. Heat transfer. In *Fish biomechanics* (ed. P.W. Webb and D. Weihs), pp. 248-279. New York: Praeger.

Graham, J.B., Dewar, H., Lai, N.C., Lowell, W.R. & Arce, S.M., 1990. Aspects of shark swimming performance determined using a large water tunnel. *Journal of Experimental Biology*, **151**, 175-192.

Graham, J.B., Gee, J.H., Motta, J. & Rubinoff, I., 1987a. Subsurface buoyancy regulation by the sea snake *Pelamis platurus*. *Physiological Zoology*, **60**, 251-261.

Graham, J.B., Gee, J.H. & Robison, F.S., 1975. Hydrostatic and gas exchange functions of the lung of the sea snake *Pelamis platurus*. *Comparative Biochemistry and Physiology*, **50A**, 477-482.

Graham, J. B., Lowell, W. R., Rubinoff, I. & Motta, J., 1987b. Surface and subsurface swimming of the sea snake *Pelamis platurus*. *Journal of Experimental Biology*, **127**, 27-44.

Grain, J., 1987. Les ciliés. In *Traité de zoologie*, vol. 2 (ed. P.P. Grassé), p. 821. Paris: Masson.

Gray, J., 1933a. Studies in animal locomotion. I. The movement of fish, with special reference to the eel. *Journal of Experimental Biology*, **10**, 88-104.

Gray, J., 1933b. Studies in animal locomotion. III. The propulsive mechanism of the whiting (*Gadus merlangus*). *Journal of Experimental Biology*, **10**, 391-400.

Greenwood, P.H., Rosen, D.E., Weitzman, S.H. & Myers, G.S., 1966. Phyletic studies of teleostean fishes, with a provisional classification of living forms. *Bulletin of the American Museum of Natural History*, **131**, 345-355.

Grillner, S., Buchanan, J.T. & Lansner, A., 1988. Simulation of the segmental burst generating network for locomotion in lamprey. *Neuroscience Letters*, **89**, 31-35.

Grillner, S. & Kashin, S., 1976. On the generation and performance of swimming in fish. In *Neural control of locomotion* (ed. R.M. Herman *et al.*), pp. 181-201. New York: Plenum.

Grillner, S., Wallén, P., Brodin, L. & Lansner, A., 1991. Neuronal network generating locomotor behavior in lamprey: circuitry, transmitters, membrane properties, and simulation. *Annual Reviews of Neuroscience*, **14**, 169-199.

Grobecker, D.B. & Pietsch, T.W., 1979. High-speed cinematographic evidence for ultrafast feeding in antennariid anglerfishes. *Science, New York*, **205**, 1161-1162.

Gross, A.C., Kyle, C.R. & Malewicki, D.J., 1983. The aerodynamics of human-powered land vehicles. *Scientific American*, **249** 6), 126-134.

Haefner, P.A. Jr, 1964. Morphometry of the common Atlantic squid, *Loligo pealei*, and the brief squid, *Lolliguncula brevis*, in Delaware Bay. *Chesapeake Science*, **5**, 138-144.

Hainsworth, F.R., 1987. Precision and dynamics of positioning by Canada geese flying in formation. *Journal of Experimental Biology*, **128**, 445-462.

Halstead, L.B., 1989. Plesiosaur locomotion. *Journal of the Geological Society, London*, **146**, 37-40.

Hanlon, R.T., Hixon, R.F., Hulet, W.H. & Yang, W.T., 1979. Rearing experiments on the California market squid, *Loligo opalescens* Berry 1911. *Veliger*, **21**, 428-431.

Harper, D.G. & Blake, R.W., 1990. Fast-start performance of rainbow trout *Salmo gairdneri* and northern pike *Esox lucius*. *Journal of Experimental Biology*, **150**, 321-342.

Harris, J.E., 1936. The role of fins in the equilibrium of swimming fish. I. Wind-tunnel tests on a model of *Mustelus canis* (Mitchell). *Journal of Experimental Biology*, **13**, 476-493.

Harris, J.E., 1937. The role of fin movements in the equilibrium of fish. *Report of the Tortugas Laboratory. Carnegie Institution*, **11**, 936-937.

Harris, J.E., 1938. The role of fins in the equilibrium of swimming fish. II. The role of the pelvic fins. *Journal of Experimental Biology*, **15**, 32-47.

Harrison, P., Nicol, J.M. & Johnston, I.A., 1987. Gross morphology, fibre composition and mechanical properties of pectoral fin muscles in the Antarctic teleost, *Notothenia neglecta* Nybelin. In *Proceedings of the Fifth Congress of European Ichthyology* (ed. K.O. Kulander and B. Fernholm), pp. 459-465. Stockholm.

Hartman, D.S., 1979. Ecology and behaviour of the manatee (*Trichechus manatus*) in Florida. *Special Publication of the American Society of Mammalogists*, **5**, 1-153.

Hartman, F.A., 1961. Locomotor mechanisms of birds. *Smithsonian Miscellaneous Collections*, **143**, 1-91.

Hauff, B., 1953. *Das Holzmadenbuch*. Öhringen.

He, P. & Wardle, C.S., 1986. Tilting behaviour of the Atlantic mackerel, *Scomber scombrus*, at low swimming speeds. *Journal of Fish Biology*, **29** (Supplement A), 223-232.

He, P. & Wardle, C.S., 1988. Endurance at intermediate swimming speeds of Atlantic mackerel, *Scomber scombrus* L., herring, *Clupea harengus* L., and saithe, *Pollachius virens* L. *Journal of Fish Biology*, **33**, 255-266.

Heatwole, H., Minton, S.A. Jr, Taylor, R. & Taylor, V., 1978. Underwater observations on sea snake behaviour. *Records of the Australian Museum*, **31**, 737-761.

Heatwole, H. & Seymour, R., 1975. Diving physiology. In *The biology of sea snakes* (ed. W.A. Dunson), pp. 289-327. Baltimore: University Park Press.

Hertel, H., 1966. *Structure - form - movement*. New York: Rheinhold.

Hess, E.H., 1959. Imprinting. *Science, New York*, **130**, 133-141.

Hess, F., 1983. Bending moments and muscle power in swimming fish. *Proceedings of the 8th Australasian Fluid Mechanics Conference*. Vol. 2, pp. 12A1-3. University of New Castle, New South Wales.

Hess, F. & Videler, J.J., 1984. Fast continuous swimming of saithe (*Pollachius virens*): a dynamic analysis of bending moments and muscle power. *Journal of Experimental Biology*, **109**, 229-251.

Higdon, J.J.L. & Corrsin, S., 1978. Induced drag of a bird flock. *American Naturalist*, **112**, 727-744.

Hildebrand, M., 1985. Walking and running. In *Functional vertebrate morphology* (ed. M. Hildebrand et al.), pp. 38-57. Cambridge, Mass.: Harvard University Press.

Hill, A.V., 1938. The heat of shortening and the dynamic constants of muscle. *Proceedings of the Royal Society of London* B, **126**, 136-195.

Hill, A.V., 1964. The efficiency of mechanical power development during muscular shortening and its relation to load. *Proceedings of the Royal Society of London* B, **159**, 319-324.

Hill, R.W., 1972. Determination of oxygen consumption by use of paramagnetic oxygen analyzer. *Journal of Applied Physiology*, **33**, 261-263.

Hoar, W.S. & Randall, D.J., 1978. *Fish physiology*. Vol. 7. *Locomotion*. New York: Academic Press.

Hochachka, P.W., French, C.J. & Meredith, J., 1978. Metabolic and ultrastructural organization in *Nautilus* muscles. *Journal of Experimental Zoology*, **205**, 51-62.

Hochachka, P.W. & Somero, G.N., 1984. *Biochemical adaptation*. Princeton University Press.

Hoerner, S.F., 1965. *Fluid-dynamic drag*. Midland Park, New Jersey: Hoerner.

Hoerner, S.F., 1975. *Fluid-dynamic lift*. Brick Town: Hoerner Fluid Dynamics.

Holland, K.N., Brill, R.W. & Chang, R.K.C., 1990. Horizontal and vertical movements of yellowfin and bigeye tuna associated with fish aggregating devices. *Fishery Bulletin. National Oceanic and Atmospheric Administration. Washington, DC*, **88**, 493-507.

Holland, K.M., Brill, R.W., Chang, R.K.C., Sibert, J.R. & Fournier, D.A., 1992. Physiological and behavioural thermoregulation in bigeye tuna (*Thunnus obesus*). *Nature, London*, **358**, 410-412.

Hollande, A., Cachon, J., Cachon, M. & Valentin, J., 1967. Infrastructure des axopodes et organisation générale de *Sticholonche zanclea* Hertwig (Radiolaire Sticholonchidea). *Protistologica*, **3**, 155-164.

Holwill, M.E.J., 1989. Biomechanical properties of the sliding filament mechanism. In *Cell movement*, vol. 1 (ed. V.I.D. Warner *et al.*), pp. 199-258. New York: Allan R. Liss Inc.

Houten, J. van, 1989. Chemoreception in unicellular eukaryotes. *Evolution of the first nervous systems* (ed. P. Anderson), pp. 343-356. New York: Plenum Press. [Life Sciences Series, 188.]

Howland, H.C., 1974. Optimal strategies for predator avoidance: the relative importance of speed and manoeuvrability. *Journal of Theoretical Biology*, **47**, 333-350.

Hoyt, D.F. & Taylor, C.R., 1981. Gait and the energetics of locomotion in horses. *Nature, London*, **292**, 239-240.

Hudson, R.C.L., 1973. On the function of the white muscles in teleosts at intermediate swimming speeds. *Journal of Experimental Biology*, **58**, 509-522.

Hui, C.A., 1983. *Swimming in penguins*. PhD thesis, University of California.

Hui, C.A., 1985. Maneuverability of the Humboldt penguin (*Spheniscus humboldti*) during swimming. *Canadian Journal of Zoology*, **63**, 2165-2167.

Hummel, D., 1983. Aerodynamic aspects of formation flight in birds. *Journal of Theoretical Biology*, **104**, 321-347.

Hustler, K., 1992. Buoyancy and its constraints on the underwater foraging behaviour of reed cormorants *Phalacrocorax africanus* and darters *Anhinga melanogaster*. *Ibis*, **134**, 229-236.

Huxley, A.F., 1957. Muscle structure and theories of contraction. *Progress in Biophysics and Biophysical Chemistry*, **7**, 255-318.

Jackson, D.C., 1969. Buoyancy control in the freshwater turtle, *Pseudemys scripta elegans*. *Science, New York*, **166**, 1649-1651.

Jacobs, D.K. & Landman, N.H., 1993. Is *Nautilus* a good model for the form and function of ammonoids? *Lethaia*, in press.

Jahn, T.L. & Bovee, E.C., 1967. Motile behavior of protozoa. *Research in protozoology*, vol. 1 (ed. Tze-Tuan Chen), pp. 42-200. Oxford: Pergamon Press.

James, R.S., Altringham, J.D. & Goldspink, D.F., 1994. The mechanical properties of fast and slow skeletal muscles of the mouse in relation to their locomotory function. *Journal of Experimental Biology*, in press.

Johnsen, S. & Kier, W.M., 1993. Intramuscular crossed connective tissue fibres: skeletal support in the lateral fins of squid and cuttlefish (Mollusca: Cephalopoda). *Journal of Zoology*, **232**, 311-338.

Johnson, T.P. & Johnston, I.A., 1991. Power output of fish muscle fibres performing oscillatory work: effects of acute and seasonal temperature change. *Journal of Experimental Biology*, **157**, 409-423.

Johnson, W., Soden, P.D. & Trueman, E.R., 1972. A study in jet propulsion: an analysis of the motion of the squid *Loligo vulgaris*. *Journal of Experimental Biology*, **56**, 155-165.

Johnston, I.A., 1981. Structure and function of fish muscle. *Symposia of the Zoological Society of London*, **48**, 71-113.

Johnston, I.A., 1991. Muscle action during locomotion: a comparative perspective. *Journal of Experimental Biology*, **160**, 167-185.

Johnston, I.A., Davison, W. & Goldspink, G., 1977. Energy metabolism of carp swimming muscles. *Journal of Comparative Physiology*, **114**, 203-216.

Johnston, I.A., Franklin, C.E. & Johnson, T.P., 1993. Recruitment patterns and contractile properties of fast muscle fibres isolated from rostral and caudal myotomes of the short-horned sculpin. *Journal of Experimental Biology*, in press.

Johnston, I.A. & Moon, T.W., 1980. Endurance exercise training in the fast and slow muscles of a teleost fish, *Pollachius virens*. *Journal of Comparative Physiology*, **135**, 147-156.

Jordan, C.E., 1992. A model of rapid-start swimming at intermediate Reynolds number: undulatory locomotion in the chaetognath *Sagitta elegans*. *Journal of Experimental Biology*, **163**, 119-137.

Josephson, R.K., 1985a. Mechanical power output from striated muscle during cyclic contraction. *Journal of Experimental Biology*, **114**, 493-512.

Josephson, R.K., 1985b. The mechanical power output of a tettigoniid wing muscle during singing and flight. *Journal of Experimental Biology*, **117**, 357-368.

Josephson, R.K. & Stokes, D.R., 1989. Strain, muscle length and work output in a crab muscle. *Journal of Experimental Biology*, **145**, 45-61.

Kaiser, H.E., 1960. Untersuchungen zur vergleichenden Osteologie der fossilen und rezenten Pachyostosen. *Palaeontographica A*, **114**, 113-196.

Keenleyside, M.H.A., 1979. *Diversity and adaptation in fish behaviour*. New York: Springer-Verlag.

Kesseler, H., 1966. Beitrag zur Kenntnis des chemischen und physicalischen Eigegenschaften des Zellsaftes von *Noctiluca miliaris*. *Veroffentlichungen des Instituts für Meeresforschung in Bremerhaven*, **2**, 357-368.

Kessler, J.O., 1985. Co-operative and concentrative phenomena of swimming microorganisms. *Contemporary Physics*, **26**, 147-166.

Kessler, J.O., 1989. Path and pattern - the mutual dynamics of swimming cells and their environment. *Comments on Theoretical Biology*, **1**, 85-108.

Kiceniuk, J.W. & Jones, D.R., 1977. The oxygen transport system in trout (*Salmo gairdneri*) during sustained exercise. *Journal of Experimental Biology*, **69**, 247-260.

Kier, W.M., 1989. The fin musculature of cuttlefish and squid (Mollusca, Cephalopoda): morphology and mechanics. *Journal of Zoology*, **217**, 23-38.

Kier, W.M., Messenger, J.B. & Miyan, J.A., 1985. Mechanoreceptors in the fins of the cuttlefish, *Sepia officinalis*. *Journal of Experimental Biology*, **119**, 369-373.

Kier, W.M., Smith, K.K. & Miyan, J.A., 1989. Electromyography of the fin musculature of the cuttlefish *Sepia officinalis*. *Journal of Experimental Biology*, **143**, 17-31.

King, A.S. & King, D.Z., 1979. Avian morphology: general principles. In *Form and function in birds*, vol. 1 (ed. A.S. King and J. McLelland) pp. 1-38. London: Academic Press.

Kishinouye, K., 1923. Contributions to the comparative study of the so-called scombroid fishes. *Journal of the College of Agriculture. Imperial University of Tokyo*, **8**, 295-473.

Kleiber, M., 1947. Body size and metabolic rate. *Physiological Reviews*, **27**, 511-541.

Klima, M., 1992. Schwimmbewegungen und Auftauchmodus bei Walen und bei Ichthyosaurien. I. Anatomische Grundlagen der Schwimmbewegungen. *Natur und Museum. Frankfurt*, **122**, 1-17.

Knower, T., Shadwick, R.E., Biewener, A.A., Korsmeyer K. & Graham, J.B., 1993a. Direct measurements of tail tendon forces in swimming tuna. *American Zoologist*, **30**, 30A.

Knower, T., Shadwick, R.E., Wardle, C.S., Korsmeyer, K. & Graham, J.B., 1993b. The timing of red muscle activation in swimming tuna. *American Zoologist*, **30**, 30A.

Kooyman, G.L., 1975. Behaviour and physiology of diving. In *The biology of penguins* (ed. B. Stonehouse), pp. 115-137. London: MacMillan.

Kopell, N. & Ermentrout, G.B., 1988. Coupled oscillators and the design of central pattern generators. *Mathematical Biosciences*, **90**, 87-109.

Korsmeyer, K.E., Graham, J.B. & Shadwick, R.E., 1992. Cardiac performance in swimming yellowfin tuna. *American Zoologist*, **32**, 57A.

Krohn, M. & Boisclair, D., 1994. The use of a stereo-video system to estimate the energy expenditure of free-swimming fish. *Canadian Journal of Fisheries and Aquatic Sciences*, in press.

Kropach, C., 1975. The yellow-bellied sea snake, *Pelamis*, in the Eastern Pacific. In *The biology of sea snakes* (ed. W.A. Dunson), pp. 185-213. Baltimore: University Park Press.

Kyle, C.R., 1979. Reduction of wind resistance and power output of racing cyclists and runners traveling in groups. *Ergonomics*, **22**, 387-397.

Lai, N.C., Graham, J.B. & Burnett, L., 1990a. Blood respiratory properties and the effect of swimming on blood gas transport in the leopard shark *Triakis semifasciata*. *Journal of Experimental Biology*, **151**, 161-173.

Lai, N.C., Graham, J.B., Lowell, W.R. & Shabetai, R., 1989. Elevated pericardial pressure and cardiac output in the leopard shark *Triakis semifasciata* during exercise: the role of the pericardioperitoneal canal. *Journal of Experimental Biology*, **147**, 263-277.

Lai, N.C., Shabetai, R., Graham, J.B., Hoit, B.D., Sunnerhagen, K. & Bhargava, V., 1990b. Cardiac function of the leopard shark, *Triakis semifasciata*. *Journal of Comparative Physiology B*, **160**, 259-268.

Lane, F.W., 1960. *Kingdom of the octopus. The life history of the Cephalopoda.* New York: Pyramid Publications.

Lang, T.G., 1975. Speed, power and drag measurements of dolphins and porpoises. In *Swimming and flying in nature*, vol. 2 (ed. T.Y.T. Wu *et al.*), pp. 553-572. New York: Plenum Press.

Lauder, G.V., 1989. Caudal fin locomotion in ray-finned fishes: historical and functional analyses. *American Zoologist*, **29**, 85-102.

Lauder, G.V. & Liem, K.F., 1983. The evolution and interrelationships of the actinopterygian fishes. *Bulletin of the Museum of Comparative Zoology at Harvard College*, **150**, 95-197.

Laval, M., 1968. *Zoothamnium pelagicum* Du Plessis, cilié pétriche planctonique: morphologie, croissance et comportement. *Protistologica*, **4**, 333-364.

Leeuwen, J.L. van, Altringham, J.D. & Johnston, I.A., 1991. Power output of fish swimming muscle: model and experiment compared. (Abstract.) *Journal of the Marine Biological Association of the United Kingdom*, **71**, 724.

Leeuwen, J.L. van, Lankheet, M.J.M., Akster, H.A. & Osse, J.W.M., 1990. Function of red axial muscle of carp (*Cyprinus carpio* L.): recruitment and normalised power output during swimming in different modes. *Journal of Zoology*, **220**, 123-145.

Legal, E. & Reichel, P., 1880. Über die Beziehung der Größe der Flugmuskulatur sowie der Größe der Flügelflächen zum Flugvermögen. *Jahresbericht der Schlesischen Gesellschaft für Vaterländische Kultur. Breslau*, **57**, 72-108, 234-270.

Lenfant, C., Johansen, K. & Torrance, J.D., 1970. Gas transport and oxygen storage capacity in some pinnipeds and the sea otter. *Respiration Physiology*, **9**, 277-286.

Levandowsky, M. & Kaneta, P.J., 1987. Behaviour in dinoflagellates. In *The biology of dinoflagellates* (ed. F.R.J. Taylor), pp. 360-397. Oxford: Blackwells. [Botanical Monographs vol. 21.]

Lighthill, M.J., 1960. Note on the swimming of slender fish. *Journal of Fluid Mechanics*, **9**, 305-317.

Lighthill, M.J., 1969. Hydrodynamics of aquatic animal propulsion. *Annual Review of Fluid Mechanics*, **1**, 413-446.

Lighthill, M.J., 1971. Large-amplitude elongated body theory of fish locomotion. *Proceedings of the Royal Society of London B*, **179**, 125-138.

Lighthill, M.J., 1973. On the Weis-Fogh mechanism of lift generation. *Journal of Fluid Mechanics*, **60**, 1-17.

Lighthill, M.J., 1975. *Mathematical biofluiddynamics.* Philadephia: Society for Industrial and Applied Mathematics.

Lighthill, M.J., 1990a. Biofluiddynamics of balistiform and gymnotiform locomotion. Part 2. The pressure distribution arising in two-dimensional irrotational flow from a general symmetrical motion of a flexible flat plate normal to itself. *Journal of Fluid Mechanics*, **213**, 1-10.

Lighthill, M.J., 1990b. Biofluiddynamics of balistiform and gymnotiform locomotion. Part 3. Momentum enhancement in the presence of a body of elliptical cross-section. *Journal of Fluid Mechanics*, **213**, 11-20.

Lighthill, M.J., 1990c. Biofluiddynamics of balistiform and gymnotiform locomotion. Part 4. Short-wavelength limitations on momentum enhancement. *Journal of Fluid Mechanics*, **213**, 21-32.

Lighthill, M.J. & Blake, R.W., 1990. Biofluid dynamics of balistiform and gymnotiform locomotion. Part 1. Biological background, and analysis by elongated-body theory. *Journal of Fluid Mechanics*, **212**, 183-207.

Lindsey, C.C., 1978. Form, function, and locomotory habits in fish. In *Fish physiology*. Vol. 7. *Locomotion* (ed. W.S. Hoar and D.J. Randall), pp. 1-100. New York: Academic Press.

Lingham-Soliar, T., 1991. Locomotion in mosasaurs. *Modern Geology*, **16**, 229-248.

Lissaman, P.B.S. & Shollenberger, C.A., 1970. Formation flight of birds. *Science, New York*, **168**, 1003-1005.

Livezey, B., 1988. Morphometrics of flightlessness in Alcidae. *Auk*, **105**, 681-698.

Livezey, B., 1989. Morphometric patterns in recent and fossil penguins (Aves, Sphenisciformes). *Journal of Zoology*, **219**, 269-307.

Lovvorn, J.R. & Jones, D.R., 1991a. Effects of body size, body fat, and change in pressure with depth on buoyancy and costs of diving in ducks (*Aythya* spp.). *Canadian Journal of Zoology*, **69**, 2879-2887.

Lovvorn, J.R. & Jones, D.R., 1991b. Body mass, volume, and buoyancy of some aquatic birds, and their relation to locomotor strategies. *Canadian Journal of Zoology*, **69**, 2888-2892.

Lowe, P.R., 1933. On the primitive characters of penguins and their bearing on the phylogeny of birds. *Proceedings of the Zoological Society of London*, **1933**, 483-538.

Lowe, P.R., 1939. Some additional notes on Miocene penguins in relation to their origin and systematics. *Ibis*, **14**, 281-294.

Machemer, H., 1974. Ciliary activity and metachronism in protozoa. In *Cilia and flagella* (ed. M. Sleigh), pp. 199-286. New York: Academic Press.

Machemer, H., 1988. Electrophysiology. In *Paramecium* (ed. H.-D. Gortz), pp. 186-215. Berlin: Springer Verlag.

Machemer, H. & Sugino, K., 1989. Electrophysiological control of ciliary beating: a basis of motile behaviour in ciliated protozoa. *Comparative Biochememistry and Physiology*, **94**, 365-374.

Machin, K.E., 1958. Wave propagation along flagella. *Journal of Experimental Biology*, **35**, 796-806.

Magnan, A., 1922. Les caractéristiques des oisseux suivant le mode de vol. *Annales des Sciences Naturelles (Zoologie)*, série 5, **10**, 125-334.

Magnuson, J.J., 1978. Locomotion by scombrid fishes: hydromechanics, morphology and behaviour. In *Fish physiology*, vol. VII (ed. W.S. Hoar and D.J. Randall), pp. 239-313. New York: Academic Press.

Manter, J.T., 1940. The mechanics of swimming in the alligator. *Journal of Experimental Zoology*, **83**, 345-358.

Marchaj, C.A., 1988. *Aero-hydrodynamics of sailing*. Camden: International Marine Publishing.

Marples, B.J., 1952. Early Tertiary penguins of New Zealand. *Palaeontological Bulletin. New Zealand Geological Survey. Wellington*, **20**, 1-66.

Marsh, R.L., 1990. Deactivation rate and shortening velocity as determinants of contractile frequency. *American Journal of Physiology*, **259**, R223-R230.

Martin, L.D. & Rothschild, B., 1989. Paleopathology and diving mosasaurs. *American Scientist*, **77**, 460-467.

Maruyama, T., 1981. Motion of the longitudinal flagellum in *Ceratium tripos* (Dinoflagellida): a retractile flagellar motion. *Journal of Protozoology*, **28**, 328-336.

Maruyama, T., 1982. Fine structure of the longitudinal flagellum in *Ceratium tripos*, a marine dinoflagellate. *Journal of Cell Science*, **58**, 109-123.

Maruyama, T., 1985. Ionic control of the longitudinal flagellum in *Ceratium tripos* (Dinoflagellida). *Journal of Protozoology*, **32**, 106-110.

Massare, J.A., 1988. Swimming capabilities of Mesozoic marine reptiles: implications for method of predation. *Paleobiology*, **14**, 187-205.

Massare, J.A., 1992. Ancient mariners. *Natural History*, **101**(9), 48-53.

Massare, J.A. & Callaway, J.M., 1990. The affinities and ecology of Triassic ichthyosaurs. *Bulletin of the Geological Society of America*, **102**, 409-416.

McCosker, J.B., 1975. Feeding behavior of Indo-Australian Hydrophiidae. In *The biology of sea snakes* (ed. W.A. Dunson), pp. 217-232. Baltimore: University Park Press.

McFadden, G.I., Schulze, D., Surek, B., Salisbury, J.L. & Melkonian, M., 1987. Basal body reorientation mediated by a Ca^{2+}-modulated contractile protein. *Journal of Cell Biology*, **105**, 903-912.

McGowan, C., 1979. A revision of the Lower Jurassic ichthyosaurs of Germany with descriptions of two new species. *Palaeontographica A*, **166**, 93-135.

McGowan, C., 1983. *The successful dragons: a natural history of Mesozoic reptiles*. Toronto: Samuel Stevens.

McGowan, C., 1989. The ichthyosaurian tailbend: a verification problem facilitated by computed tomography. *Paleobiology*, **15**, 429-436.

McGowan, C., 1990. Computed tomography confirms that *Eurhinosaurus* (Reptilia: Ichthyosauria) does have a tailbend. *Canadian Journal of Earth Science*, **21**, 1541-1545.

McGowan, C., 1992. The ichthyosaurian tail: sharks may not provide an appropriate analogue. *Palaeontology*, **35**, 555-570.

McLaughlin, R.L. & Kramer, D.L., 1991. The association between amount of red muscle and mobility in fishes: a statistical evaluation. *Environmental Biology of Fishes*, **30**, 369-378.

Meckel, J.F., 1828. *System der vergleichenden Anatomie*. Band 3, pp. 286-392. Halle.

Menzbier, M., 1887. Vergleichende Osteologie der Pinguine in Anwendung zur Haupteinteilung der Vögel. *Bulletin de la Société Impériale des Naturalistes de Moscou*, **3**, 483-587.

Merriam, J.C., 1908. Triassic ichthyosaurs with special reference to the American forms. *Memoirs of the University of California*, **1**, 1-96.

Meunier, K., 1958. Die Allometrie des Vogelflügels. *Zietschrift für Wissenschaftliche Zoologie*, **161**, 444-482.

Mill, G.K. & Baldwin, J., 1983. Biochemical correlates of swimming and diving behaviour in the little penguin *Eudyptula minor*. *Physiological Zoology*, **56**, 242-254.

Minasian, S.M., Balcomb, K.C. III & Foster, L., 1984. *The world's whales*. Washington, DC: Smithsonian Books.

Montgomery, J.C. & Macdonald, J.A., 1984. Performance of motor systems in Antarctic fishes. *Journal of Comparative Physiology*, **154A**, 241-248.

Moon, T.W., Altringham, J.D. & Johnston, I.A., 1991. Energetics and power output of isolated fish fast muscle fibres performing oscillatory work. *Journal of Experimental Biology*, **158**, 261-273.

Mordvinov, Ju.E., 1980. Hydrodynamics of the rockhopper penguin. *Biologiya Morya*, **5**, 52-56. [In Russian.]

Moss, M.L., 1977. Skeletal tissues in sharks. *American Zoologist*, **17**, 335-342.

Moy-Thomas, J.A. & Miles, R.S., 1971. *Palaeozoic fishes*. Philadelphia: Saunders Company.

Muelder, E.W. & Theunissen, B., 1986. Herman Schlegel's investigation of the Maastricht mosasaurs. *Archives of Natural History*, **13**, 1-6.

Murata, M., 1988. On the flying behaviour of neon flying squid *Ommastrephes bartrami* observed in the central and northwestern north Pacific. *Nippon Suisan Gakkaishi*, **54**, 1167-1174.

Myers, E.H., 1943. Biology, ecology and morphogenesis of a pelagic foraminifer. *Stanford University Publications. Biological Sciences*, **9**, 1-30.

Nachtigall, W. & Bilo, O.D., 1980. Strömungsanpassung des Pinguins beim Schwimmen unter Wasser. *Journal of Comparative Physiology*, **137A**, 17-26.

Nachtigall, W. & Klimbingat, A., 1985. Messungen der Flügelgeometrie mit der Profilkamm-Methode und geometrische Flügelkennzeichnung einheimischer Eulen. In *BIONA-report 3* (ed. W. Nachtigall), pp. 45-86. Akadamia Wissenschaftliche Mainz: G. Fischer.

Nachtigall, W. & Wieser, J., 1966. Profilmessungen am Taubenflügel. *Zeitschrift für Vergleichende Physiologie*, **52**, 333-346.

Naef, A., 1921/23. Cephalopoda. *Fauna e flora del Golfo di Napoli*. Monograph no. 35, Part I, Fascicle I.

Naitoh, Y. & Sugino, K., 1984. Ciliary movement and its control in *Paramecium*. *Journal of Protozoology*, **31**, 31-40.

Nicholls, E.L. & Russell, A.P., 1991. The plesiosaur pectoral girdle: the case for a sternum. *Neues Jahrbuch für Geologie und Paläontologie, Abhandlungen*, **182**, 161-185.

Nishiwaki, M. & Marsh, H., 1985. Dugong *Dugong dugon* (Müller, 1776). In *Handbook of marine mammals.* Vol. 3. *The sirenians and baleen whales* (ed. S.H. Ridgway and R.J. Harrison), pp. 1-31. London: Academic Press.

Norberg, U.M., 1970. Functional osteology and myology of the wing of *Plectotus auritus* Linnaeus (Chiroptera). *Arkiv för Zoologi,* **22**, 483-543.

Norberg, U.M., 1972. Functional osteology and myology of the wing of the dog-faced bat *Rousettus aegyptiacus* (E. Geoffry). *Zeitschrift für Morphologie der Tiere,* **73**, 1-44.

Norberg, U.M., 1990. *Vertebrate flight.* New York: Springer-Verlag.

Norberg, U.M. & Rayner, J.M.V., 1987. Ecological morphology and flight in bats (Mammalia; Chiroptera): wing adaptations, flight performance, foraging strategy and echolocation. *Philosophical Transactions of the Royal Society of London* B, **316**, 335-427.

Norris, K.S. & Prescott, J.H., 1961. Observations on Pacific cetaceans of Californian and Mexican waters. *University of California Publications in Zoology,* **63**, 291-402.

O'Dor, R.K., 1982. Respiratory metabolism and swimming performance of the squid, *Loligo opalescens. Canadian Journal of Fisheries and Aquatic Sciences,* **39**, 580-587.

O'Dor, R.K., 1988a. The energetic limits on squid distributions. *Malacologia,* **29**, 113-119.

O'Dor, R.K. 1988b. The forces acting on swimming squid. *Journal of Experimental Biology,* **137**, 421-442.

O'Dor, R.K. & Balch, N., 1985. Properties of *Illex illecebrosus* egg masses potentially influencing larval oceanographic distribution. *Scientific Council Studies. Northwest Atlantic Fisheries Organization,* **9**, 69-76.

O'Dor, R.K., Carey, F.G., Webber, D.M. & Voegeli, F.M., 1991. Behaviour and energetics of Azorean squid, *Loligo forbesi.* In *Proceedings of the 11th International Symposium on Biotelemetry* (ed. C.J. Amlaner), pp. 191-195. Fayetteville: University of Arkansas Press.

O'Dor, R.K., Forsythe, J., Webber, D.M., Wells, J. & Wells, M.J., 1993a. Activity levels of *Nautilus* in the wild. *Nature, London,* **362**, 626-628.

O'Dor, R.K., Foy, E.A., Helm, P.L. & Balch, N., 1986. The locomotion and energetics of hatchling squid, *Illex illecebrosus. American Malacological Bulletin,* **4** (1), 55-60.

O'Dor, R.K., Hoar, J.A., Webber, D.M., Carey, F.G., Tanaka, S., Martins, H. & Porteiro, F.M., 1993b. Squid (*Loligo forbesi*) performance and metabolic rates in nature. *Marine Behavior and Physiology,* in press.

O'Dor, R.K. & Webber, D.M., 1986. The constraints on cephalopods: why squid aren't fish. *Canadian Journal of Zoology,* **64**, 1591-1605.

O'Dor, R.K. & Webber, D.M., 1991. Invertebrate athletes: trade-offs between transport efficiency and power density in cephalopod evolution. *Journal of Experimental Biology,* **160**, 93-112.

O'Dor, R.K., Wells, J. & Wells, M.J., 1990. Speed, jet pressure and oxygen consumption relationships in free-swimming *Nautilus. Journal of Experimental Biology,* **154**, 383-396.

Oehme, H., 1970. Vergleichende Profiluntersuchungen an Vogelflügeln. *Beiträge zur Vogelkunde. Leipzig,* **16**, 301-312.

Oehme, H. & Bannasch, R., 1989. Energetics of locomotion in penguins. In *Energy transformation in cells and organisms* (ed. W. Wieser and E. Gnaiger), pp. 230-240. Stuttgart: Thieme.

Oehme, H., Dathe, H.H. & Kitzler, U., 1977. Flight energetics in birds. Research on biophysics and physiology of the bird flight IV. *Fortschritter der Zoologie. Jena,* **24**, 257-273.

Orthmeyer, D.L. & Ball, I.J., 1990. Survival of mallard broods on Benton Lake National Wildlife Refuge in northcentral Montana. *Journal of Wildlife Management,* **54**, 62-66.

Otis, T.S. & Gilly, W.F., 1990. Jet-propelled escape in the squid *Loligo opalescens*: concerted control by giant and non-giant motor axon pathways. *Proceedings of the National Academy of Sciences of the United States of America,* **87**, 2911-2915.

Owen, R., 1836. *Aves. Cyclopedia of anatomy and physiology.* Volume 1, pp. 265-358. London: Sherwood, Gilbert & Piper.

Owen, R., 1866. *Comparative anatomy and physiology of vertebrates.* Volume 2. London: Longmans, Green & Co.

Packard, A., 1966. Operational convergence between cephalopods and fish: an exercise in functional anatomy. *Archivio Zoologico Italiano,* **51**, 523-542.

Packard, A., 1969. Jet propulsion and the giant fibre response of *Loligo*. *Nature, London*, **221**, 875-877.

Packard, A., 1972. Cephalopods and fish: the limits of convergence. *Biological Reviews*, **47**, 241-307.

Packard, A., Bone, Q. & Hignette, M., 1980. Breathing and swimming movements in a captive *Nautilus*. *Journal of the Marine Biological Association of the United Kingdom*, **60**, 313-327.

Partridge, B.L. & Pitcher, T.J., 1979. Evidence against hydrodynamic function for fish schools. *Nature, London*, **279**, 418-419.

Patterson, C., 1968a. The caudal skeleton in lower Liassic pholidophoroid fishes. *Bulletin of the British Museum (Natural History). Geology*, **16**, 203-239.

Patterson, C., 1968b. The caudal skeleton in Mesozoic acanthopterygian fishes. *Bulletin of the British Museum. (Natural History). Geology*, **17**, 49-102.

Pedley, T.J. & Kessler, J.O., 1990. A new continuum model for suspensions of gyrotactic micro-organisms. *Journal of Fluid Mechanics*, **212**, 155-82.

Pedley, T.J. & Kessler, J.O., 1992. Hydrodynamic phenomena in suspensions of swimming micro-organisms. *Annual Review of Fluid Mechanics*, **24**, 313-58.

Pennycuick, C.J., 1972. Soaring behaviour and performance of some East African birds, observed from a motor-glider. *Ibis*, **141**, 178-218.

Pfennig, N., 1962. Beobachtungen über das Schwärmen von *Chromatium okenii*. *Archive für Mikrobiologie*, **42**, 90-95.

Piiper, J., Meyer, M., Worth, H. & Willmer, H., 1977. Respiration and circulation during swimming activity in the dogfish *Scyliorhinus stellaris*. *Respiration Physiology*, **30**, 221-239.

Pinna, G. & Nosotti, S., 1989. Anatomia, morfologia funzionale e paleoecologia del rettile placodonta *Psephoderma alpinum* Meyer, 1858. *Memorie della Società Italiana di Scienze Naturali e del Museo Civico di Storia Naturale di Milano*, **25**, 17-50.

Pond, C.M., 1978. Morphological aspects and the ecological and mechanical consequences of fat deposition in wild vertebrates. *Biological Reviews*, **9**, 519-570.

Prandtl, L., 1935. *Ergebnisse der Aerodynamischen Versuchsanstalt zu Göttingen*. 1. Lieferung 4. Auflage. Oldenbourgh, München und Berlin.

Prandtl, L. & Betz, A., 1935. *Ergebnisse der Aerodynamischen Versuchanstalt zu Goettingen*. 3. Lieferung 2. Auflage. Oldenbourgh, München und Berlin.

Prandtl, L. & Tietjens, O.G., 1934. *Applied hydro and aerodynamics*. New York: Dover.

Prange, H.D. & Schmidt-Nielsen, K., 1970. The metabolic cost of swimming in ducks. *Journal of Experimental Biology*, **53**, 763-777.

Pugh, L.G.C.E., 1971. The influence of wind resistance in running and walking and the mechanical efficiency of work against horizontal or vertical forces. *Journal of Physiology*, **213**, 255-276.

Pycraft, W.P., 1898. Contribution to the osteology of birds, part 2. Impennes. *Proceedings of the Zoological Society of London*, 1878, 958-989.

Pycraft, W.P., 1907. On some points in the anatomy of the emperor and adelie penguins. In *National Antarctic Expedition 1901-1904. Natural History*. Volume 2, *Vertebrata*, pp. 1-28. London: British Museum (Natural History).

Rayner, J.M.V., 1985. Vorticity and propulsion mechanics in swimming and flying animals. In *Konstruktionsprinzipien lebender und ausgestorbener Reptilien* (ed. J. Riess and E. Frey), pp. 89-118. Universität Stuttgart und Tübingen. [Konzepte SFB 230, 4.]

Rayner, J.M.V., Jones, G., & Thomas, A., 1986. Vortex flow visualizations reveal change in upstroke function with flight speed in bats. *Nature, London*, **321**, 162-164.

Rayner, M.D. & Keenan, M.J., 1967. Role of red and white muscles in the swimming of skipjack tuna. *Nature, London*, **214**, 392-393.

Reid, J., 1835. Anatomical description of the Patagonian penguin. *Proceedings of the Zoological Society of London*, **3**, 132-148.

Riess, J., 1984. How to reconstruct paleoecology - outlines of a holistic view and an introduction to ichthyosaur biomechanics. In *Third symposium on Mesozoic terrestrial ecosystems* (ed. W.-E. Reif and F. Westphal), pp. 201-205. Tübingen: Attempto Verlag.

Riess, J., 1986. Fortbewegungsweise, Schwimmbiophysik und Phylogenie der Ichthyosaurier. *Palaeontographica* A, **192**, 93-155.

Riess, J. & Frey, E., 1991. The evolution of underwater flight and the locomotion of plesiosaurs. In *Biomechanics in evolution* (ed. J.M.V. Rayner and R.J. Wootton), pp. 131-144. Cambridge University Press. [*Society for Experimental Biology Seminar Series*, 36.]

Roberts, A.M., 1975. In *Swimming and flying in nature*, vol. 1 (ed. T.Y.T. Wu *et al.*), pp. 377-393. New York: Plenum Press.

Roberts, A.M., 1981. Hydrodynamics of protozoan swimming. In *Biochemistry and physiology of Protozoa*, vol. 4 (ed. M. Levandowsky and S.H. Hutner), pp. 5-66. New York: Academic Press.

Roberts, J.L. & Graham, J.B., 1979. Effect of swimming speed on the excess temperatures and activities of heart and red and white muscles in the mackerel, *Scomber japonicus*. *Fishery Bulletin. National Oceanic and Atmospheric Administration. Washington DC*, **76**, 861-867.

Robinson, J.A., 1975. The locomotion of plesiosaurs. *Neues Jahrbuch für Geologie und Paläontologie, Abhandlungen*, **149**, 286-332.

Robinson, J.A., 1977. Intracorporal force transmission in plesiosaurs. *Neues Jahrbuch für Geologie und Paläontologie, Abhandlungen*, **153**, 86-128.

Rome, L.C. & Kushmerick, M.J., 1983. The energetic cost of generating isometric force as a function of temperature in isolated frog muscle. *American Journal of Physiology*, **244**, C100-C109.

Rome, L.C. & Sosnicki, A.A., 1990. The influence of temperature on mechanics of red muscle in carp. *Journal of Physiology*, **427**, 151-169.

Rome, L.C. & Sosnicki, A.A., 1991. Myofilament overlap in swimming carp. II. Sarcomere length changes during swimming. *American Journal of Physiology*, **260**, C289-C296.

Rome, L.C. & Swank, D., 1992. The influence of temperature on power output of scup red muscle during cyclical length changes. *Journal of Experimental Biology*, **171**, 261-281.

Rome, L., Swank, D. & Corda, D., 1993. How fish power swimming. *Science, New York*, **261**, 340-343.

Rome, L.C., 1990. The influence of temperature on muscle recruitment and function *in vivo*. *American Journal of Physiology*, **259**, R210-R222.

Rome, L.C., Choi, I-H., Lutz, G. & Sosnicki, A.A., 1992a. The influence of temperature on muscle function in the fast swimming scup. I. Shortening velocity and muscle recruitment during swimming. *Journal of Experimental Biology*, **163**, 259-279.

Rome, L.C., Funke, R.P. & Alexander, R.M., 1990. The influence of temperature on muscle velocity and sustained performance in swimming carp. *Journal of Experimental Biology*, **154**, 163-178.

Rome, L.C., Funke, R.P., Alexander, R.McN., Lutz, G., Aldridge, H.D.J.N., Scott, F. & Freadman, M., 1988. Why animals have different muscle fibre types. *Nature, London*, **335**, 824-827.

Rome, L.C., Loughna, P.T. & Goldspink, G., 1984. Muscle fiber recruitment as a function of swim speed and muscle temperature in carp. *American Journal of Physiology*, **247**, R272-R279.

Rome, L.C., Loughna, P.T. & Goldspink, G., 1985. Temperature acclimation improves sustained swimming performance at low temperatures in carp. *Science, New York*, **228**, 194-196.

Rome, L.C., Sosnicki, A.A. & Choi, I-H., 1992b. The influence of temperature on muscle function in the fast swimming scup. II. The mechanics of red muscle. *Journal of Experimental Biology*, **163**, 281-295.

Rome, L.C., Swank, D. & Corda, D., 1993. How fish power swimming. *Science, New York*, **261**, 340-343.

Romer, A.S. & Parson, T.S., 1977. *The vertebrate body*. Philadelphia: Saunders Company.

Roper, C.F.E. & Boss, K.J., 1982. The giant squid. *Scientific American*, **246**(4), 82-89.

Rosen, D.E., 1982. Teleostean interrelationships, morphological function and evolutionary inference. *American Zoologist*, **22**, 261-273.

Rothschild, B. & Martin, L.D., 1987. Avascular necrosis: occurrence in diving Cretaceous mosasaurs. *Science, New York*, **236**, 75-77.

Round, F.E. & Crawford, R.M., 1990. Phylum Bacillariophyta. In *Handbook of Protoctista* (ed. L. Margulis *et al.*), pp. 574-596. Boston: Jones & Bartlett.

Russell, D.A., 1967. Systematics and morphology of American mosasaurs. *Peabody Museum of Natural History, Yale University, Bulletin*, **23**, 1-240.

Salisbury, J.L., 1983. Contractile flagellar roots: the role of calcium. *Journal of Submicroscopic Cytology*, **15**, 105-110.

Salisbury, J.L., 1989. Centrin and the algal flagellar apparatus. *Journal of Phycology*, **25**, 201-206.

Salisbury, J.L., Baron, A., Surek, B. & Melkonian, M., 1984. Striated flagellar roots: isolation and partial characterization of a calcium-modulated contractile organelle. *Journal of Cell Biology*, **99**, 962-970.

Salisbury, J.L. & Floyd, G.L., 1978. Calcium-induced contraction of the rhizoplast of a quadriflagellate green alga. *Science, New York*, **202**, 975-977.

Satchell, G.H., 1991. *Physiology and form of fish circulation*. Cambridge: Cambridge University Press.

Satir, P., 1989. Structural analysis of the dynein cross-bridge cycle. In *Cell movement*, vol. 1 (ed. V.I.D. Warner *et al.*), pp. 219-234. New York: Allan R. Liss Inc.

Schaeffer, B., 1967. Osteichthyan vertebrae. *Journal of the Linnean Society of London (Zoology)*, **47**, 185-195.

Scharold, J., Lai, N.C., Lowell, W.R. & Graham, J.B., 1989. Metabolic rate, heart rate, and tailbeat frequency during sustained swimming in the leopard shark *Triakis semifasciata*. *Experimental Biology*, **48**, 223-230.

Schliwa, M., Shimizu, T., Vale, R.D. & Euteneur, U., 1991. Nucleotide specificities of anterograde and retrograde organelle transport in *Reticulomyxa* are indistinguishable. *Journal of Cell Biology*, **112**, 1199-1203.

Schmidt-Nielsen, K., 1972. Locomotion: energy cost of swimming, flying and running. *Science, New York*, **177**, 222- 228.

Schmidt-Nielsen, K., 1977. Problems of scaling: locomotion and physiological correlates. In *Scale effects in animal locomotion* (ed. T.J. Pedley), pp 1-21. London: Academic Press.

Schoenlein, K., 1894. Beobachtungen über blutkreislauf und respiration bein einigen fischen. *Zeitschrift für Biologie*, **32**, 511-547.

Schoepss, C.G., 1829. Beschreibung der Flügelmuskeln der Vögel. *Meckels Archiv für Anatomie und Physiologie (Leipzig)*, **1829**, 72-176.

Schreiweis, D.O., 1982. A comparative study of the appendicular musculature of penguins (Aves: Sphenisciformes). *Smithsonian Contributions to Zoology*, **341**, 1-46.

Segawa, S., Yang, W.T., Marthy, H.-J. & Hanlon, R.T., 1988. Illustrated embryonic stages of the eastern Atlantic squid *Loligo forbesi*. *Veliger*, **30**, 230-243.

Shabetai, R., Abel, D.C., Graham, J.B., Bhargava, V., Keyes, R.S. & Witztum, K., 1985. Function of the pericardium and the pericardioperitoneal canal in elasmobranch fishes. *American Journal of Physiology*, **248**, H198-H207.

Shapere, A. & Wilczek, F., 1989a. Geometry of self-propulsion at low Reynolds number. *Journal of Fluid Mechanics*, **198**, 557-585.

Shapere, A. & Wilczek. F., 1989b. Efficiency of self-propulsion at low Reynolds number. *Journal of Fluid Mechanics*, **198**, 587-598.

Shaw, E., 1978. Schooling fishes. *American Scientist*, **66**, 166-175.

Shufeldt, R.W., 1907. Osteology of the penguins. *Journal of Anatomy and Physiology. London*, **35**, 390-404.

Sibley, C.G., 1991. Phylogeny and classification of birds from DNA comparisons. *Acta XX Congressus Intenationalis Ornithologici. Programme and abstracts*, pp. 111-126. Wellington: New Zealand Ornithological Congress.

Sigvardt, K.A. & Williams, T.L., 1992. Models of central pattern generators as oscillators: mathematical analysis and simulations of the lamprey locomotor CPG. *Seminars in the Neurosciences*, **4**, 37-46.

Sim, E., 1986. *Fin thrust production and fin activity in* Loligo pealei. BSc thesis, Dalhousie University, Halifax, Canada.

Simpson, G.G., 1946. Fossil penguins. *Bulletin of the American Museum of Natural History*, **87**, 1-100.

Simpson, G.G., 1975. Fossil penguins. In *The biology of penguins* (ed. B. Stonehouse), pp 19-42. London: MacMillan.

Sleigh, M.A., 1984. The integrated activity of cilia: function and coordination. *Journal of Protozoology*, **31**, 16-21.

Sleigh, M.A., 1989a. Ciliary propulsion in protozoa. *Scientific Progress. Oxford*, **73**, 317-332.

Sleigh, M.A., 1989b. Adaptations of ciliary systems for the propulsion of water and mucus. *Comparative Biochemistry and Physiology*, **94**, 359-364.

Sliter, W.V., 1965. Laboratory experiments on the life cycle and ecological controls of *Rosalina globularis* d'Orbigny. *Journal of Protozoology*, **12**, 210-215.

Sornette, D., 1989. A physical model of cell crawling motion. *Journal de Physiologie, France*, **50**, 1759-1770.

Sosnicki, A.A., Loesser, K. & Rome, L.C., 1991. Myofilament overlap in swimming carp. I. Myofilament lengths of red and white muscle. *American Journal of Physiology*, **260**, C283-C288.

Sournia, A., 1986. *Atlas du phytoplancton marin.* Vol. 1. Paris: CNRS.

Soyer, M.O., 1970. Les ultrastructures liées aux fonctions de relation chez *Noctiluca miliaris* (Dinoflagellata). *Zeitschrift für Zellforschung und Mikroskopische Anatomie*, **104**, 29-55.

Stegman, B., 1937. Über die Flügelhaltung von Archaeornis in der Ruhestellung. *Ornithologische Monatsberichte. Berlin*, **45**, 192-195.

Stegman, B., 1970. Peculiarities of morphology of the wing in penguins. *Trudy Zoologicheskogo Instituta Akademiya Nauk*, **47**, 236 - 248. [In Russian.]

Stephan, B., 1979. Vergleichende Osteologie der Pinguine. *Mitteilungen aus dem Zoologischen Museum in Berlin*, **55**, Suppl. Ann. Ornithol, 3-98.

Stephens, R.E. & Stommel, E.W., 1989. Role of cAMP in ciliary and flagellar motility in cell movement. In *Cell movement*, vol. 1 (ed. V.I.D. Warner *et al.*), pp. 299-316. New York: Allan R. Liss Inc.

Stephenson, R., Lovvorn, J.R., Heieis, M.R.A., Jones, D.R. & Blake, R.W., 1989. A hydromechanical estimate of the power requirements of diving and surface swimming in lesser scaup (*Aythya affinis*). *Journal of Experimental Biology*, **147**, 507-518.

Stevenson, R.D. & Josephson, R.K., 1990. Effects of operating frequency and temperature on mechanical power output from moth flight muscle. *Journal of Experimental Biology*, **149**, 61-78.

Stolpe, M. & Zimmer, K., 1939. Der Schwirrflug des Kolobri im Zeitlupenfilm. *Journal für Ornithologie*, **87**, 136-155.

Stonehouse, B., 1975. Introduction. In *The biology of penguins* (ed. B. Stonehouse), pp 1-16. London: MacMillan.

Storer, R.W., 1960. Evolution in the diving birds. *International Ornithological Congress*, **12**, 694-707.

Storrs, G.W., 1993. Function and phylogeny in sauropterygian (Diapsida) evolution. *American Journal of Science*, **293A**, 63-90.

Sy, M.H., 1936. Funktionell-anatomische Untersuchungen am Vogelflügel. *Journal für Ornithologie*, **84**, 199-196.

Symmons, S., 1979. Notochordal and elastic components of the axial skeleton of fishes and their functions in locomotion. *Journal of Zoology*, **189**, 157-206.

Tang, J. & Wardle, C.S., 1992. Power output of two sizes of Atlantic salmon (*Salmo salar*) at their maximum sustained swimming speeds. *Journal of Experimental Biology*, **166**, 33-46.

Tarsitano, S. & Riess, J., 1982. Plesiosaur locomotion - underwater flight versus rowing. *Neues Jahrbuch für Geologie und Paläontologie, Abhandlungen*, **164**, 188-192.

Taylor, F.J.R., 1987. Dinoflagellate morphology. In *Biology of dinoflagellates* (ed. F.R.J. Taylor), pp. 24-91. Oxford: Blackwells. [Botanical Monographs vol. 21.]

Taylor, G., 1952. Analysis of the swimming of long and narrow animals. *Proceedings of the Royal Society of London A*, **214**, 158-183.

Taylor, M.A., 1981. Plesiosaurs - rigging and ballasting. *Nature, London*, **290**, 628-629.

Taylor, M.A., 1987. A reinterpretation of ichthyosaur swimming and buoyancy. *Palaeontology*, **30**, 531-535.

Taylor, M.A., 1991. Stone, bone or blubber? Buoyancy control strategies in aquatic tetrapods. *Journal of the Marine Biological Association of the United Kingdom*, **71**, 722-723. [Abstract.]

Taylor, M.A., 1993. Stomach stones for feeding or buoyancy? The occurrence and function of gastroliths in marine tetrapods. *Philosophical Transactions of the Royal Society of London B*, **341**, 163-175.

Thomson, K.S., 1976. On the heterocercal tail in sharks. *Palaeobiology*, **2**, 19-38.

Thomson, K.S. & Simanek, E.D., 1977. Body form and locomotion in sharks. *American Zoologist*, **17**, 343-345.

Trueman, E.R., 1975. *The locomotion of soft-bodied animals*. New York: American Elsevier Publishing Co. Inc.

Trueman, E.R. & Packard, A., 1968. Motor performances of some cephalopods. *Journal of Experimental Biology*, **49**, 495-507.

Tsukamoto, K., 1981. Direct evidence for functional and metabolic differences between dark and ordinary muscle in free-swimming yellowtail, *Seriola quinqueradiata*. *Bulletin of the Japanese Society of Scientific Fisheries*, **47**, 573-575.

Tsukamoto, K., 1984. The role of the red and white muscles during swimming of the yellowtail. *Bulletin of the Japanese Society of Scientific Fisheries*, **50**, 2025-2030.

Tucker, V.A., 1975. The energetic cost of moving about. *American Scientist*, **63**, 413-419.

Vaughn, P.P. & Dawson, M.R., 1956. On the occurrence of calcified tympanic membranes in the mosasaur *Platecarpus*. *Transactions of the Kansas Academy of Science*, **59**, 382-384.

Videler, J.J., 1977. Mechanical properties of fish tail joints. *Fortschritte der Zoologie*, **24**, 183-194.

Videler, J.J., 1981. Swimming movements, body structure and propulsion in cod *Gadus morhua*. *Symposia of the Zoological Society of London*, **48**, 1-27.

Videler, J.J., 1993. *Fish Swimming*. London: Chapman & Hall. [Fish and Fisheries Series 10.]

Videler, J.J. & Hess, F., 1984. Fast continuous swimming of two pelagic predators: saithe (*Pollachius virens*) and mackerel (*Scomber scombrus*): a kinematic analysis. *Journal of Experimental Biology*, **109**, 209-228.

Videler, J.J. & Wardle, C.S., 1978. New kinematic data from high speed cine film recordings of swimming cod (*Gadus morhua*). *Netherlands Journal of Zoology*, **28**, 465-484.

Videler, J.J. & Wardle, C.S., 1991. Fish swimming stride by stride: speed limits and endurance. *Reviews in Fish Biology and Fisheries*, **1**, 23-40.

Videler, J.J. & Weihs, D., 1982. Energetic advantages of burst-and-coast swimming of a fish at high speeds. *Journal of Experimental Biology*, **97**, 169-178.

Vigues, B., David, C. & Bayle, D., 1992. Ciliary zones retraction in entodiniomorphid ciliates: possible involvement of a calcium modulated contractile protein. *Cell Motility and the Cytoskeleton*, **23**, 311.

Vlymen, W.J., 1974. Swimming energetics of larval anchovy, *Engraulis mordax*. *Fishery Bulletin. National Oceanographic and Atmospheric Administration. Washington DC*, **72**, 885-899.

Vogel, S., 1981. *Life in moving fluids: the physical biology of flow*. Boston: Willard Grant Press.

Vogel, S. & LaBarbera, M., 1978. Simple flow tanks for research and teaching. *Bioscience*, **28**, 638-643.

Wade, M., 1984. *Platypterygius australis*, an Australian Cretaceous ichthyosaur. *Lethaia*, **17**, 99-113.

Wade, M., 1990. A review of the Australian Cretaceous longipinnate ichthyosaur *Platypterygius* (Ichthyosauria, Ichthyopterygia). *Memoirs of the Queensland Museum*, **28**, 115-137.

Wainwright, S.A., Biggs, J.D., Currey, J.D. & Gosline, J.M., 1976. *Mechanical design in organisms*. London: Edward Arnold.

Wallén, P. & Williams, T.L., 1984. Fictive locomotion in the lamprey spinal cord *in vitro* compared with swimming in the intact and spinal animal. *Journal of Physiology*, **347**, 225-239.

Walls, W.P., 1983. The correlation between high limb-bone density and aquatic habits in recent mammals. *Journal of Paleontology*, **57**, 197-207.

Wardle, C.S., 1975. Limit of fish swimming speed. *Nature, London*, **255**, 725-727.

Wardle, C.S. & Videler, J.J., 1980. How do fish break the speed limit? *Nature, London*, **284**, 445-447.

Wardle, C.S. & Videler, J.J., 1993. The timing of the electromyogram in the lateral myotomes of mackerel and saithe at different swimming speeds. *Journal of Fish Biology*, **42**, 347-359.

Wardle, C.S., Videler, J.J., Arimoto, T., Franco, J.M. & He, P., 1989. The muscle twitch and the maximum swimming speed of the giant bluefin tuna, *Thunnus thynnus* L. *Journal of Fish Biology*, **35**, 129-137.

Wassersug, R.J., 1989. Locomotion in amphibian larvae (or 'Why aren't tadpoles built like fishes?'). *American Zoologist*, **29**, 65-84.

Watson, D.M.S., 1924. The elasmosaurid shoulder-girdle and forelimb. *Proceedings of the Zoological Society of London*, **1924**, 885-917.

Watson, M., 1883. Report on the anatomy of the Spheniscidae collected during the voyage of HMS Challenger. *Challenger Report, Zooology*, **7** (18), 1-242.

Weast, R.C., Astle, M.J. & Beyer, W.H. (ed.), 1988. *CRC handbook of chemistry and physics*, 69th edition. Boca Raton, Florida: CRC Press.

Webb, P.W., 1971. The swimming energetics of trout. I. Thrust and power output at cruising speeds. *Journal of Experimental Biology*, **55**, 489-520.

Webb, P.W., 1975a. Hydrodynamics and energetics of fish propulsion. *Bulletin of the Fisheries Research Board of Canada*, **190**, 1-159.

Webb, P.W., 1975b. Efficiency of pectoral-fin propulsion of *Cymatogaster aggregata*. In *Swimming and flying in nature*, vol. 2 (ed. T.Y.T. Wu *et al.*), pp. 573-584. New York: Plenum Press.

Webb, P.W., 1975c. Acceleration performance of rainbow trout *Salmo gairdneri* and green sunfish *Lepomis cyanellus*. *Journal of Experimental Biology*, **63**, 451-465.

Webb, P.W., 1976. The effect of size on the fast-start performance of rainbow trout *Salmo gairdneri*, and a consideration of piscivorous predator-prey interactions. *Journal of Experimental Biology*, 65, 157-177.

Webb, P.W., 1978a. Fast-start performance and body form in seven species of teleost fish. *Journal of Experimental Biology*, **74**, 211-226.

Webb, P.W., 1978b. Temperature effects on acceleration of rainbow trout, *Salmo gairdneri*. *Journal of the Fisheries Research Board of Canada*, **35**, 1417-1422.

Webb, P.W., 1978c. Hydrodynamics: non-scombroid fish. In *Fish Physiology*, vol. 7 (ed. W.S. Hoar and D.J. Randall), pp 189-237. New York: Academic Press.

Webb, P.W., 1982. Locomotor patterns in the evolution of actinopterygian fishes. *American Zoologist*, **22**, 329-342.

Webb, P.W., 1983. Speed, acceleration and manoeuvrability of two teleost fishes. *Journal of Experimental Biology*, **102**, 115-122.

Webb, P.W., 1984. Body form, locomotion and foraging in aquatic vertebrates. *American Zoologist*, **24**, 107-120.

Webb, P.W., 1986. Locomotion and predator-prey relationships. In *Predator-prey relationships*, (ed. M.E. Feder and G. V. Lauder), pp. 24-41. Chicago University Press.

Webb, P.W., 1988. Simple physical principles and vertebrate aquatic locomotion. *American Zoologist*, **28**, 709-725.

Webb, P.W., 1989. Station-holding by three species of benthic fishes. *Journal of Experimental Biology*, **145**, 303-320.

Webb, P.W., 1990. How does benthic living affect body volume, tissue composition, and density of fishes? *Canadian Journal of Zoology*, **68**, 1250-1255.

Webb, P.W., 1991. Composition and mechanics of routine swimming of rainbow trout, *Oncorhynchus mykiss*. *Canadian Journal of Fisheries and Aquatic Sciences*, **48**, 583-590.

Webb, P.W., 1992. Is the high cost of body/caudal fin undulatory swimming due to increased friction drag or inertial recoil? *Journal of Experimental Biology*, **162**, 157-166.

Webb, P.W., 1993. Swimming. In *Fish physiology* (ed. D.H. Evans), pp. 47-73. New York: CRC Press.

Webb, P.W., 1994a. Exercise performance of fish. In *Comparative vertebrate exercise physiology* (ed. J.H. Jones), in press. Orlando: Academic Press.

Webb, P.W., 1994b. Is tilting behaviour at low swimming speeds unique to negatively buoyant fish? Observations on steelhead trout, *Oncorhynchus mykiss*, and bluegill, *Lepomis macrochirus*. *Journal of Fish Biology*, **43**, 687-694.

Webb, P.W. & Blake, R.W., 1985. Swimming. In *Functional vertebrate morphology* (ed. M. Hildebrand *et al.*), pp. 110-128. Cambridge, Mass.: Harvard University Press.

Webb, P.W. & Buffrenil, V. de, 1990. Locomotion in the biology of large aquatic vertebrates. *Transactions of the American Fisheries Society*, **119**, 629-641.

Webb, P.W., Hardy, D.H. & Mehl, V.L., 1992. The effect of armored skin on the swimming of longnose gar, *Lepisosteus osseus*. *Canadian Journal of Zoology*, **70**, 1173-1179.

Webb, P.W. & Keyes, R.S., 1981. Division of labour between median fins in swimming dolphin (Pisces: Coryphaeidae). *Copeia*, **1981**, 901-904.

Webb, P.W., Kostecki, P.T. & Stevens, E.D., 1984. The effect of size and swimming speed on locomotion kinematics of rainbow trout. *Journal of Experimental Biology*, **109**, 77-95.

Webb, P.W. & Skadsen, J.M., 1979. Reduced skin mass: an adaptation for acceleration in some teleost fishes. *Canadian Journal of Zoology*, **57**, 1570-1575.

Webb, P.W. & Smith, G.R., 1980. Function of the caudal fin in early fishes. *Copeia*, **1980**, 559-562.

Webb, P.W. & Weihs, D. (ed.), 1983. *Fish biomechanics*. New York: Praeger.

Webb, P.W. & Weihs, D., 1986. Functional locomotor morphology of early life history stages of fishes. *Transactions of the American Fisheries Society*, **115**, 115-127.

Webb, P.W. & Zhang, H., 1994. The effect of changes in the responsiveness of heat-shocked goldfish on their success in avoiding attacks by rainbow trout. *Canadian Journal of Zoology*, in press.

Webber, D.M., 1985. *Monitoring the metabolic rate and activity of* Illex illecebrosus *with telemetered jet pressures*. MSc thesis, Dalhousie University, Halifax, NS, Canada.

Webber, D.M. & O'Dor, R.K., 1985. Respiration and swimming performance of short-finned squid (*Illex illecebrosus*). *Scientific Council Studies. Northwest Atlantic Fisheries Organization*, **9**, 133-138.

Webber, D.M. & O'Dor, R.K., 1986. Monitoring the metabolic rate and activity of free-swimming squid with telemetered jet pressure. *Journal of Experimental Biology*, **126**, 205-224.

Weihs, D., 1972. A hydrodynamic analysis of fish turning manoeuvres. *Proceedings of the Royal Society of London* B, **182**, 59-72.

Weihs, D., 1972. Semi-infinite vortex trails, and their relation to oscillating airfoils. *Journal of Fluid Mechanics*, **54**, 679-690.

Weihs, D., 1973. Hydromechanics of fish schooling. *Nature, London*, 241, 290-291.

Weihs, D., 1973a. Mechanically efficient swimming techniques for fish with negative buoyancy. *Journal of Marine Research*, **31**, 194-209.

Weihs, D., 1973b. The mechanism of rapid starting of slender fish. *Biorheology*, **10**, 343-350.

Weihs, D., 1974. The energetic advantages of burst swimming. *Journal of Theoretical Biology*, **49**, 215-229.

Weihs, D., 1975. Some hydrodynamical aspects of fish schooling. In *Swimming and flying in nature*, vol. 2 (ed. T.Y.T. Wu, C.J. Brokaw and C. Brennen), pp. 703-717. New York: Plenum Press.

Weihs, D., 1977. Effects of size on sustained swimming speeds of aquatic organisms. In *Scale effects in animal locomotion* (ed. T.J. Pedley), pp. 333-338. New York: Academic Press.

Weihs, D., 1978. Tidal stream transport as an efficient method for migration. *Journal du Conseil Permanent International pour l'Exploration de la Mer*, **38**, 92-99.

Weihs, D., 1980. Energetic significance of changes in swimming modes during growth of anchovy larvae, *Engraulis mordax*. *Fishery Bulletin. National Oceanographic and Atmospheric Administration. Washington DC*, **77**, 597-604.

Weihs, D., 1981. Effects of swimming path curvature on the energetics of fish motion. *Fishery Bulletin. National Oceanographic and Atmospheric Administration. Washington DC*, **79**, 171-176.

Weihs, D., 1989. Design features and mechanics of axial locomotion in fish. *American Zoologist*, **29**, 151-160.

Weihs, D. & Webb, P.W., 1983. Optimization of locomotion. In *Fish biomechanics* (ed. P.W. Webb and D. Weihs), pp. 339-375. New York: Praeger.

Weis-Fogh, T., 1973. Quick estimates of flight fitness in hovering animals, including novel mechanisms for lift production. *Journal of Experimental Biology*, **59**, 169-230.

Wells, M.J., 1990. Oxygen extraction and jet propulsion in cephalopods. *Canadian Journal of Zoology*, **68**, 815-824.

Wells, M.J. & O'Dor, R.K., 1991. Jet propulsion and the evolution of cephalopods. *Bulletin of Marine Science*, **49**, 419-432.

Wells, M.J. & Wells, J., 1985. Ventilation and oxygen uptake by *Nautilus*. *Journal of Experimental Biology*, **118**, 297-312.

Werner, E.E., 1986. Species interactions in freshwater fish communities. In *Community ecology* (ed. J. Diamond and T.J. Case), pp. 344-357. New York: Harper and Row.

Williams, T.L., 1991. The neural-mechanical link in lamprey locomotion. In *Neural and sensory mechanisms in locomotion* (ed. D.M. Armstrong and B.M.H. Bush), pp. 183-195. Manchester University Press.

Williams, T.L., 1992. Phase coupling by synaptic spread in chains of coupled neuronal oscillators. *Science, New York*, **258**, 662-665.

Williams, T.L., Grillner, S., Smoljaninov, V.V., Wallen, P., Kashin, S. & Rossignol, S., 1989. Locomotion in lamprey and trout: the relative timing of activation and movement. *Journal of Experimental Biology*, **143**, 559-566.

Williams, T.L., Sigvardt, K.A., Kopell, N., Ermentrout, G.B. & Remler, M.P., 1990. Forcing of coupled nonlinear oscillators: studies of intersegmental co-ordination in the lamprey locomotor central pattern generator. *Journal of Neurophysiology*, **64**, 862-871.

Williams, T.M., 1983. Locomotion in the North American mink, a semi aquatic mammal. I. Swimming energetics and body drag. *Journal of Experimental Biology*, **103**, 155-168.

Williams, T.M., 1989. Swimming by sea otters: adaptations for low energetic cost locomotion. *Journal of Comparative Physiology*, **164**A, 815-824.

Williams, T.M., Friedl, W.A., Fong, M.L., Yamada, R.M., Sedivy, P. & Haun, J.E., 1992. Travel at low energetic cost by swimming and wave-riding bottlenose dolphins. *Nature, London*, **355**, 821-823.

Williamson, G.R., 1965. Underwater observations of the squid *Illex illecebrosus* Lesueur in Newfoundland waters. *Canadian Field Naturalist*, **79**, 239-247.

Williston, S.W., 1898. Mosasaurs. *University Geological Survey of Kansas*, **4**, 83-221.

Wilson, R.P., Hustler, K., Ryan, P.G., Burger, A.E. & Nöldeke, E.C., 1992. Diving birds in cold water: do Archimedes and Boyle determine energetic costs? *American Naturalist*, **140**, 179-200.

Wiman, C., 1905. Über die alttertiären Vertebraten der Seymourinsel. *Wissenschaftliche Ergebnisse der Schwedischen Südpolar-Expedition, 1901-1903*, **3**, 1-37.

Winemiller, K.O., 1991. Ecomorphological diversification in lowland freshwater fish assemblages from five biotic regions. *Ecological Monographs*, **61**, 343-365.

Witman, G.B., 1990. Introduction to cilia and flagella. *Ciliary and flagellar membranes*, vol. 1 (ed. R.A. Bloodgood), pp. 1-30. Plenum Publishing Corporation.

Woakes, A.J. & Butler, P.J., 1983. Swimming and diving in tufted ducks, *Aythya fuligula*, with particular reference to heart rate and gas exchange. *Journal of Experimental Biology*, **107**, 311-329.

Woledge, R.C., 1992. Relaxation as a determinant of the locomotory role of muscle. *Society for Experimental Biology. Annual Meeting.* [Abstract.]

Wootton, R.J., 1990. *Ecology of teleost fishes*. London: Chapman and Hall.

Wu, T.Y., 1977. Introduction to the scaling of aquatic animal locomotion. In *Scale effects in animal locomotion* (ed. T.J. Pedley), pp. 203-232. New York: Academic Press.

Yates, G.T., 1983. Hydromechanics of body and caudal fin propulsion. In *Fish biomechanics* (ed. P.W. Webb and D. Weihs), pp. 177-213. New York: Praeger.

Zuev, G.V., 1964. The body shape of Cephalopoda. *Trudy Sevastopol Skoi Biologicheskoi Stantsii. Akademiya Nauk SSSR*, **17**, 379-387. [National Institute of Oceanography Translation, 1965.]

Zuev, G.V., 1965a. Concerning the mechanisms involved in the creation of a lifting force by the bodies of cephalopod molluscs. *Biofizika, Akademiya Nauk SSSR*, **10**, 360-361. [Canadian Fisheries and Marine Services Translation, series 999, 1968.]

Zuev, G.V., 1965b. The functional basis of the structure of the locomotory apparatus in the phylogeny of cephalopods. [*Zhurnal Obshchei Bioloii. Moskva*, **26**, 616-619. Canadian Fisheries and Marine Services Translation, series 1000, 1968.]

Zuev, G.V., 1966. Characteristic features of the structure of cephalopod molluscs associated with controlled movements. *Ekologo-Morfologicheskie Issledovaniya Nektonnykh Zhivotnykh*. Kiev, Special Publication. [Canadian Fisheries and Marine Services Translation series 1011, 1968.]

Zug, G.R., 1971. Buoyancy, locomotion, morphology of the pelvic girdle and hindlimb, and systematics of cryptodiran turtles. *Miscellaneous Publications of the Museum of Zoology, University of Michigan*, **142**, 1-98.

Index

page